BestMasters

Springer awards "BestMasters" to the best application-oriented master's theses, which were completed at renowned chairs of economic sciences in Germany, Austria, and Switzerland in 2013.

The works received highest marks and were recommended for publication by supervisors. As a rule, they show a high degree of application orientation and deal with current issues from different fields of economics.

The series addresses practitioners as well as scientists and offers guidance for early stage researchers.

Renard Teipelke

The Thika Highway Improvement Project

Changes in the Peri-Urban Northern Nairobi Metropolitan Region

Foreword by Prof. Dr. Susanne Heeg
and Dr. Veit Bachmann

Renard Teipelke
Frankfurt, Germany

ISBN 978-3-658-04538-8   ISBN 978-3-658-04539-5 (eBook)
DOI 10.1007/978-3-658-04539-5

The Deutsche Nationalbibliothek lists this publication in the Deutsche Nationalbibliografie; detailed bibliographic data are available in the Internet at http://dnb.d-nb.de.

Library of Congress Control Number: 2013955623

Springer Gabler
© Springer Fachmedien Wiesbaden 2014

Printed on acid-free paper

Springer Gabler is a brand of Springer DE.
Springer DE is part of Springer Science+Business Media.
www.springer-gabler.de

Foreword

A wide variety of recent research on urban development in the Global South shows how a high-capacity infrastructure is of utmost importance to sustainable development of metropolitan areas. However, most analysis – both in the Global South or North – is often preoccupied with technical aspects; it commonly focuses on questions of how to make infrastructure projects feasible and technically safe, how to develop sound financial schemes or how to guarantee political support and balance different demands. Renard Teipelke's work is different in this respect: It is not primarily concerned with technical questions but embeds the infrastructure project in the particular context of the Kenyan state and includes a wider contextualization within Kenya's contemporary political society, urban situation, and developmental ambitions.

In so doing, Renard Teipelke explores backgrounds, preconditions, developments, and effects of the Thika Highway Improvement Project (THIP) – a vast road project in Kenya's capital city of Nairobi. He explores the outcomes of this megaproject in the Northern Nairobi Metropolitan Region; providing an insightful account of the project's social dimension and its meaning to the people involved in and directly affected by it. Throughout the analysis, Teipelke throws new light on the THIP, often with surprising elements. He shows how, in the dominant perception, the THIP avoids many of the problems associated with other megaprojects, such as challenges by the local population, funding shortfalls, or environmental degradation. Instead, the THIP is widely supported by the public, the private sector, and politicians. Critical voices therefore evolve less around the project in its entirety but more around the way it has been implemented and realized. In a kaleidoscopic picture of Kenyan society, Teipelke pays attention to often overlooked details. He illustrates how, due to unsettled tenure questions, the compensation for people without formal property rights to their homes or businesses is socially extremely unjust. Tenants, home owners, or entrepreneurs without a formal title, constituting the majority of inhabitants of the land used for the THIP, have not received compensation for their lost property. As such the THIP is contributing to increased social inequality, favoring those with formal land titles (and the political elite) and discriminating against the vast majority of people living and working in informal arrangements.

Surprisingly it is thereby not the project itself that is questioned but the way how social questions and legitimate demands are handled. His empirical data shows how, generally, Kenyan society is particularly sensitive with respect to questions of social involvement and political participation. However, the analysis also shows how this sensitivity is not visible in governmental strategies. This is only one of the examples where Teipelke's work goes beyond the social aspects of the THIP. Embedded in an informative contextualization with the two overarching development strategies, Kenya Vision 2030 and Nairobi Metro 2030, Teipelke provides insights into the political structure of the Kenyan state and relations between political (and business) elites and society. His work explores possible ways for overcoming the shortcomings and pitfalls of such megaprojects, and their underlying development strategies, in the Global South. It is particularly in this sense that we expect Renard Teipelke's work to find broad readership not only amongst researchers and students, but also amongst policy-makers, NGOs, and international development organizations.

Prof. Dr. Susanne Heeg and Dr. Veit Bachmann, Department of Human Geography, Goethe University Frankfurt

Acknowledgements

After having worked eight months on preparing, conducting, and finalizing a research project, I can finally understand why scholars in various kinds of publications write such emotional acknowledgements – this master thesis has been a very big step in my academic and personal development process. It was particularly this step from behind the desk into the field that brought tremendous challenges I had to cope with. And I could not have coped with them if I could not have relied on support from very special people both in Kenya and Germany.

The initial phase of a research project is often one of the hardest. While my supervisors, Susanne Heeg and Veit Bachmann, have provided invaluable advice during the entire process of the project, they were most important before the actual research work even started. It looks a bit ridiculous now that I took approximately as much time to develop the research topic and conceptual framework as I needed to conduct the field work. However, my supervisors were completely right in rebutting several suggestions early on and to pushing me into different directions to critically question assumptions and the feasibility of project ideas, before I was actually given the green light to start my field work. My supervisors' flexibility was also crucial with respect to my tight schedule and stressful, although structured, working style. While my own deadline for finishing this project was 'ambitious' to say the least, my supervisors had no doubt that I would be able to handle the situation, and their regular encouragement and critical feedback made this research project possible.

On the practical side of the field work and early conceptionalization of my research work, I am grateful to my internship's hosting office: the Urban Planning and Design Branch of UN-Habitat. My colleagues gave me important tips on how to best conduct the field work and they helped me get into contact with the relevant experts in Kenya. Furthermore, they provided me with a support letter that opened many doors in various institutions. Without the flexible work time arrangements at the end of my internship, I also could have not conducted the field work and the interviews. The Urban Planning and Design Branch's help made the empirical work of this research project feasible.

For my time in Kenya, a couple of people need to be mentioned for their input into my research work (and beyond). I want to thank Dorothee von Brentano not only for the material she contributed to my research sources, but particularly for the blunt conversations we had about Kenyan politics and planning. While many views exist on this topic, Dorothee's experience after decades of dealing with Kenyan politicians, government officials, private sector actors, and civil society organizations opened my eyes to better 'read' what actually is going on in the country. Another person who supported me and raised my curiosity from the early days on in Kenya is Simon Kinyanjui. What he used to call 'intellectual dates' became the hardest thought and discussion work that I enjoyed during my stay in Kenya. Having an opinion, fighting for an argument, and asking the right questions is our approach to critical thinking. Thus, we both gained much out of our intense conversations about politics and culture in different contexts. I also want to thank Jacqueline Klopp who has been contributing important research to this thesis' topic in recent years and was a reference point for me to understand the various aspects of infrastructure projects. I really appreciated her help in exchanging information and ideas on this topic.

Then, of course, I want to thank all the interviewees and conversation partners I talked to in order to conduct my research work. While their names remain anonymous, everyone who received a digital copy of this master thesis can be ensured that this thank goes to them, as research work like this would not be possible if experts would not be willing to invest some of their precious time to talk to researchers. I really hope that they find this thesis enriching and that they feel encouraged to continue seeking the exchange between research and practice. For their support in providing valuable material to my research, I would also like to acknowledge the Institute of Surveyors of Kenya, other institutions and organizations, as well as researchers and practitioners who made available documents that are otherwise hard to access.

I would like to apologize to the four men in the matatu from Ruiru to Nairobi who stole my wallet in a wonderful choreographed act, because that day I had so little money with me that they could hardly balance their transport costs for going back and forth. I encourage everyone who witnessed the theft to stand up more courageously against those people who incessantly work on preserving the somewhat true image of 'Nairobbery'. Besides this one incident, I was received very hospitably by Kenyans who were wonderful hosts and who provided interesting stories during my field work.

Concluding the acknowledgements for my research work in Kenya, I am greatly indebted to Thâmara Fortes for her constant motivation, emotional support, and critical feedback on my daily work in the field. Without her, I would have become desperate when dealing with the different communication culture and the unbelievably terrible traffic in the Nairobi Metropolitan Region. While expert input and professional advice were important for this research project, Thâmara's personal commitment cannot be valued highly enough.

With regard to the support I received in Germany, I would like to thank Stefan Ouma and Marc Boeckler for our intensive conversations about this research project and conclusions drawn on my field work experience. I am grateful for the opportunity I had to present parts of my master thesis at the AKSA meeting in Bayreuth, where colleagues also provided relevant input into my work. My home institution, the Department of Human Geography at Goethe University Frankfurt, was definitely the right place not only to pursue a master's degree in a cutting-edge critical research program, but also to widen my horizons and to turn to the Global South and metropolitan regions in particular, as they are my future commitment.

This master thesis would have looked very different without the critical review and helpful feedback by Caspar Lundsgaard-Hansen, Alexander Niedziolka, Emily Achenbach, Amad Judeh, and Dan Orbeck, as well as Raphael Schwegmann, Ann-Christin Hayk, Christian Girmann, Sebastian Krumpfe, Christian Posselt, and Grace Sullivan. They are all young scholars/practitioners whom I could entrust to prevent me from getting stuck in this research project without the necessary questioning of the quality of my work.

Last but not least, I am forever grateful to my parents, my sister, her partner, my niece, and my grandmother, as well as my best friends for supporting me during this process in one way or the other (yes, a case of Club Mate can be counted under this category) and who are standing behind me no matter what I am pursuing and where I am going. With the eventual super-motivational kick by the following Mercator Fellowship for International Affairs, this thesis was always on schedule and paved the way to now practically apply this research knowledge in the upcoming work placements.

Renard Teipelke

Table of Contents

List of Figures

List of Photos

1. Introduction

> *"This super-highway is a great example of (...) our commitment to transform Kenya into a strong economic hub for the region and beyond."*
> *(Kibaki 2012: 1)*

Photo 1.1: Kenya's President Mwai Kibaki (2002-2013) officially opens Thika Highway with Chinese Ambassador Liu Guanyuan on 9 November 2012 (Source: Xinhua 2012, http://www.the-star.co.ke/ sites/default/files/styles/node_article/public/images/articles/2012/11/12/95071/kenyaspresident.png).

1.1 A Milestone of Kenya's Development in the Eyes of the Kenyan People

On 9 November 2012, Kenyans were excited to see their country's first 'superhighway' being officially opened by then President Mwai Kibaki. They saw this infrastructure project as another big step of their nation towards becoming a middle-income country by 2030. In the weeks around this special day, the media was full of enthusiastic news pieces on the first eight-lane international trunk road in East Africa (for TV see NTV 2012 or Otieno 2012; for newspaper see Barasa 2012 or Odipo 2012). Even though there have been critical reports already before the construction (Wairimu 2012c; Standard 2012), the predominant feelings in the early days of November 2012 were that of joy and pride. A large infrastructure project was delivered...one year beyond schedule and with a budget of 31 instead of the originally estimated 24 billion Kenyan Shillings (370 instead of 287 million US-Dollars). But people in the streets were happy with a project that was reasonably good for what they were referring to as 'Kenyan standards', particularly, because the country has seen many 'white elephants' – projects that have been pompously promised but never delivered.

Zooming in from the outside, the Thika Highway Improvement Project (THIP) or the 'Thika Superhighway' is a large transportation infrastructure project covering the approximately 50 kilometers between Kenya's capital Nairobi and the industrial satellite town of Thika in the Northern part of the Nairobi Metropolitan Region with its 6.7 million inhabitants (Consulting Engineering Services and Runji & Partners 2011: 6-7). The project was divided into three sections (see figure 1.1): The first, most urban one was supported financially (and with regard to consultancy) by the African Development Fund of the African Development Bank Group. The other two sections were paid for by the Export-Import Bank of China. The Kenyan Government also contributed to the project budget. Construction for the three sections was undertaken separately by three Chinese companies: Shengli Engineering Construction Group Co. Ltd., Sinohydro Corporation Ltd., and China Wu Yi Company Ltd. (KARA & CSUD 2012: 6).

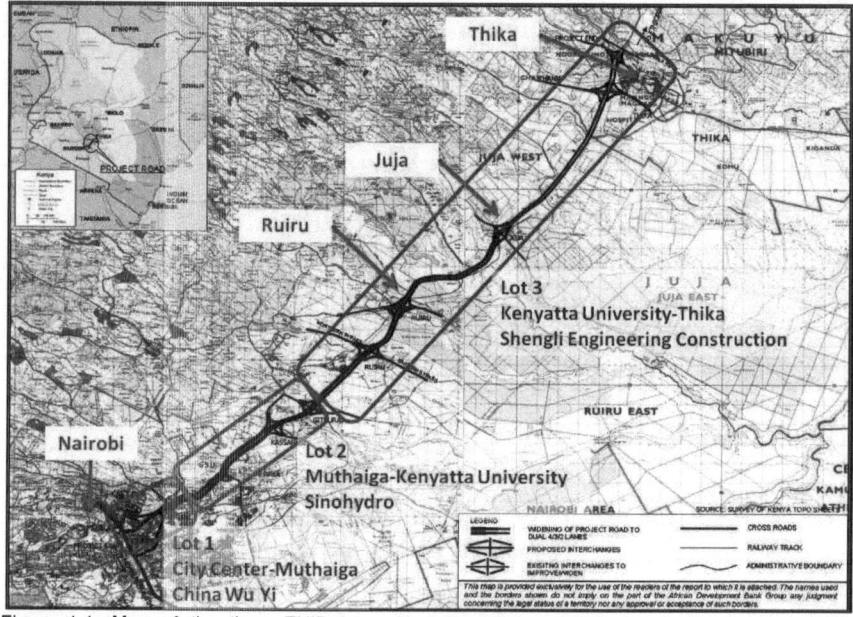

Figure 1.1: Map of the three THIP lots with information on their construction area and the implementing Chinese engineering companies (Source: ADF 2007: 21; red rectangles and text added by author).

In summary, the project's main focus lay on the expansion of the former dual carriageway (four-lane) Thika Road into a highway with six to eight lanes and dual service lanes to each side in most sections of the highway. Since this major transport corridor into and out of Nairobi was plagued with massive traffic jams, high numbers of (fatal) accidents, and a terrible traffic management, the old Thika Road was considered one of the worst transport corridors in Kenya (ADF 2007: vii; JICA 2006: 9-10). The results of the construction, which took place from the end of 2008 until the

2

end of 2012, were believed to address these and other challenges that will be discussed further in this thesis.

It is important to underscore that the THIP can be seen under the country's broad development initiative "Kenya Vision 2030" and its development strategy for the Nairobi Metropolitan Region "Nairobi Metro 2030" (GOK 2007, 2008b). In these government documents, the THIP is part of a larger package of infrastructure projects, including several bypasses for the Nairobi Metropolitan Region.

1.2   Research Perspectives for Analyzing the THIP

While the THIP has been described in public as 'one of its kind', researchers have studied roads, highways, or large transportation infrastructure projects in general for decades. In contrast to previous studies of such infrastructure projects, this thesis aims to comprehensively study the THIP and its outcomes by applying five research perspectives that shed light on the road itself, the land adjacent to it, and the life and work of people along that transportation corridor.

The first theoretical perspective applied in this thesis, road infrastructure economics, has been occupied with studying cost-benefit analyses of road projects and figuring out the relation (and causation) between infrastructure investments, economic growth, and urban as well as social development. The literature shows rather indefinite results in this regard, often depending on many contextual features of specific case studies as well as the applied econometric calculus (Boarnet & Chalermpong 2001; Sanchez 2000; Ayogu 2007; Banerjee et al. 2012). This research field has experienced a slight shift in focus when it turned to countries of the Global South and discovered the relevance of large transportation infrastructure projects in connecting 'underdeveloped' countries/regions with the aim of triggering their economic potentials to evolve to 'full potential'. Transportation corridors became the catchphrase in many donor-driven development programs and related studies – on the African continent particularly since the 21st century.

Having rather been a research niche for long, the past two decades saw the rise of transport mobility studies, which provide the second research perspective in the analytical framework of this thesis. It turns to the study of the relation between transport mobility and urban form. A much deeper understanding of the strong interdependencies of these two elements has been gained by studying road infrastructure projects from the standpoint of transportation planning in relation to urban planning (Banister 2012; UN-Habitat 2013; Todes 2012a, b). Despite a strong emphasis on mobility in research and practice, its counterpart, accessibility, is of even greater importance for the analysis in this thesis.[1] The reason for this lies in the characteristics of the THIP: The first two sections of it are going through an urban area that is still part of the city of Nairobi, while the third section – starting at Kenyatta University, passing through Ruiru and going up to the town of Thika – covers an area that can be best described as peri-urban (see figure 1.1). With cities in highly diverse countries of the Global South experiencing an extensive growth in population and a mushrooming spatial extension, the phenomenon of peri-urbanization has increasingly gained momentum in both academia and practice (Mbiba & Huchzermeyer 2002; UN-Habitat 2005; Kreibich & Olima 2002; McGregor et al.

---

[1] Key concepts such as mobility and accessibility will be discussed and defined in the theoretical chapter of this thesis.

2006). Therefore, it provides the third research perspective on studying large infrastructure projects in this thesis.

In contrast to the traditional and misguided dichotomy between 'the urban' and 'the rural', peri-urban areas are seen as transition zones in spatial, economic, social, political, ecological, and cultural terms. The understanding is still limited on how the otherwise called 'urban hinterland', the 'urban fringe', or the 'outskirts' of a metropolitan region function, what they are, what challenges they are facing, and how these challenges can be addressed best. Together with the partly limited understanding of the relation between road projects and spatial development, the area between Ruiru and Thika of the THIP becomes a prime research object, which can enrich the knowledge of both: large transportation infrastructure projects and peri-urban areas, particularly with regard to land and housing markets.

Nevertheless, it requires a broader take on Kenya's post-/colonial history and politics to fully grasp the roots of the THIP and to undercover the underlying reasons leading to the outcomes of this large infrastructure project. This task can be fulfilled by applying the rather Global North-focused urban studies literature in the context of the Global South – an amalgamation that has been convincingly called for by various critical scholars in urban and development studies (Robinson 2006; Grant 2009; Pieterse 2010b) and that provides the fourth research perspective for the analytical framework of this thesis. This means that the world/global/globalizing city literature can enhance the understanding of the envisioned development path of Kenya and Nairobi, in particular, if questions of inequality, spatial segregation, political representation, and informality are discussed. In addition, city-regions and regional governance literature – as another element of this research perspective (Ward & Jonas 2004; Harrison 2007, 2010; Jonas 2012; Neumann & Hull 2009) – is of relevance for the case study area, because it covers the Northern Nairobi Metropolitan Region which goes far beyond the administrative boundaries of the city.

Since issues of regional governance are interlinked with regional planning, the fifth research perspective on the THIP is that of planning literature. Here, it is particularly an understanding of the political-historical elements that have been influencing how politics and planning work in Kenya. The (dys-) functional role of an inherited modernist planning system can best be understood by taking a (re-) politicized view on it to see how planning relates to infrastructure projects such as the THIP and people's daily lives along transport corridors (K'Akumu & Olima 2007; Mwangi 2002; Olima 2001, 2002; UN-Habitat 2009; El-Shakhs 1997; Parnell & Simon 2010).

These five different research perspectives form the basis for the empirical analysis of the THIP and will be discussed in detail in the theoretical chapter of this thesis. This analysis will open the 'black box' THIP and uncover critical aspects in its idea, design, and implementation as the underlying elements of the project's outcomes, which are the major focus in the empirical part of this study. Since the THIP was not created in an empty space, the discussion of this project's context will also enrich the analysis. Kenya's broader development framework will be presented in order to show how the THIP emerged from a distinct political-historical background. Joining the theoretical perspectives with these insights of 'the Kenyan state', this thesis will argue that the project's outcomes are ambivalent – partly living up to the expectations of the THIP's proponents, but not resulting in shared benefits for all concerned stakeholders due to flaws in managing the development in the peri-urban transport corridor area, where the project was implemented. Essentially, this analysis will uncover the embeddedness of the THIP in the Kenyan context of politics and planning and, thereby, contribute to a more fine-grained understanding of large transportation infrastructure projects.

1.3 The Relation between Disciplines as Research Motivation

Having already introduced the 'common people' perception of the THIP as well as the theoretical perspectives on and the Kenyan context of this project, the author's own background and research motivation need to be clarified. Like scholars before me, I have conducted and experienced the empirical research work in the field in a situation of multiple personalities: I was working for the United Nations Human Settlement Program (UN-Habitat) in Nairobi for six months, during which I also discovered the country personally. Furthermore, I have studied the THIP from the position of a researcher in human geography and a social scientist in general.

As the methodological chapters will emphasize, it is essential to reflect on this, since I was dealing with the topic of large transportation infrastructure projects and the relation between different academic and professional disciplines in my practical work as well as my research work. In early November 2012, I experienced the excitement in Nairobi about the first 'superhighway' in East Africa – and questions arose like: How are 'good' infrastructure projects done? What does a city like Nairobi need in terms of infrastructure? What is known about the impact of large transportation infrastructure projects on peri-urban areas, land management, and people's livelihood? And how does the land 'out there' beyond Ruiru actually look like?

In order to answer these questions, a visit to the field was necessary. Therefore, I got onto a matatu (typical minibus in Kenya) and headed north to the Thika Highway and the towns of Ruiru, Juja, Thika, and beyond. As the elaboration at various points in this thesis will show, these field visits revealed interesting insights about the THIP, the peri-urban corridor area, and the people living and working there – insights that were not sufficiently discussed in the media coverage or the small talks on the streets of Nairobi.

But there is another aspect here: I have been educated first as a political scientist and then as a human geographer, and I was working in urban development and planning. The question which puzzled me was why experts in my research and work fields were talking about inclusive growth, just sustainability, or non-motorized transport, while nearly everyone in Kenya was jubilating about this massive new highway that physically cut through the landscape of the Northern Nairobi Metropolitan Region? What is it about 'roads' that the research community in the Global North might have misunderstood or insufficiently paid attention to in their research? Why is road infrastructure still of eminent importance in Kenya and other countries in the Global South?

This discussion can be put on a meta-level when the relation – or more often: missing link – between disciplines is studied. In their provocative article, Phelps and Tewdwr-Jones (2008) discuss this topic for the disciplines of planning and geography. They are drawing the picture of two disciplines that logically have much in common (for instance, research topics or spatialized perspectives on broader social science issues). In contrast to this 'logical' commonality, these two disciplines, planning and geography, have nevertheless existed side-by-side mostly ignoring the potentials of a fruitful interaction and despising the other discipline's style of doing research and/or engaging with practice.

Turning to the planning discipline, Phelps and Tewdwr-Jones (ibid.: 566) highlight its reluctance to engage with geographers' spatio-political analysis of shared topics. Conversely, researchers in geography are said to shy away from engaging with the 'apolitical' technical planning profession in practice. The authors explain this practicality of planning through demands in the discipline's professional activities that have necessitated a stretching of the original planning field into many other neighboring topics and professions. While this has ensured an ongoing engagement in public policy, planners have also received much criticism for what they were said to have 'planned' (ibid.: 569). In contrast, geographers were keen to preserve their critical distance from 'dirty' politics, which in turn has made their discipline more elitist and theoretical (ibid.: 576-577).

Phelps and Tewdwr-Jones (ibid.: 579) recommend that the planning discipline need to embrace more of the social science perspectives of geography, while geographic research has to accept that practical engagement will not meet their high perfectionist standards for 'good' politics and policies. A first improvement would be an intensified interaction between the disciplines that is not hindered by differences in terminologies or methods solely for the sake of the respective discipline's purity. This thesis seeks this challenge and goes one step further by juxtaposing the five above-mentioned research perspectives (road infrastructure economics, transport mobility/urban form, peri-urbanization, urban studies, and planning) with the case study of the THIP and its outcomes in the Northern Nairobi Metropolitan Region.

1.4   Outline of the Thesis

As a foundation to the analysis, chapter 2 introduces the different theoretical perspectives on the research object. After an overview, literature from road infrastructure economics is presented, followed by a discussion of literature on transport mobility and urban form. Peri-urbanization research is introduced with a focus on land and housing markets. Input from urban studies literature will be applied in the context of the Global South, before the role of planning is discussed. The theoretical chapter closes with a summary of these theoretical perspectives.

Chapter 3 is dedicated to the THIP's background. A political-historical view on the Kenyan state is offered and Nairobi as a globalizing city with its urbanization challenges is presented. Then, Kenya Vision 2030 and Nairobi Metro 2030 are introduced. This is followed by a presentation of the THIP and its context, as well as an illustration of the peri-urban study area in the Northern Nairobi Metropolitan Region.

Chapter 4 outlines the research design for the analysis of the THIP. After a formulation of the hypotheses following from the conceptual framework of this thesis, the methodology is presented. First, methodological approaches are summarized. Then, the preparation, conduction, and analysis of qualitative interviews are discussed. The final part of this chapter refers to the challenge of data in research and practice in Sub-Saharan Africa.

Chapter 5 presents the findings of the empirical analysis. In the first part, the idea of the THIP is discussed. In the second part, its design with respect to actors and processes is further analyzed. In the third part, the implementation of the THIP is scrutinized. The fourth part presents outcomes of the THIP with regard to changes to transport and economy, as well as changes to land and housing. The findings of the empirical analysis are then summarized in a conclusion.

Chapter 6 of this thesis discusses recommendations following from the theoretical and empirical analyses. First, policy recommendations are presented, followed by recommendations for researchers studying the THIP, large transportation infrastructure projects in general, as well as peri-urban areas and urbanization challenges of metropolitan regions in the Global South.

In chapter 7, the final part of this thesis, an outlook on the future of large transportation infrastructure projects in Kenya vis-à-vis recent political developments is given, followed by the bibliography and an extensive appendix with additional material for further research on the topics discussed in this thesis.

2. Theoretical Perspectives

> *"It can take a lifetime to learn about one place – how much more difficult to build analyses through a range of different contexts!" (Robinson 2006: 167)*

Photo 2.1: Thika Highway between Juja and Thika (Source: Author).

The theoretical foundation of this master thesis is built from multiple perspectives that can enrich the understanding of the various aspects of large transportation infrastructure projects. The following discussion of relevant research literature will illustrate how these perspectives emerge from different research disciplines and theoretical schools. At the same time, it will be shown how they all can be put in direct relation to the topic of this thesis – by considering the interrelation between different perspectives, identifying gaps in the existing research, and developing questions out of this theoretical discussion.

First, research from road infrastructure economics will be introduced, followed by a discussion on the relation between transport mobility and urban form. The research field of peri-urbanization will be presented with regard to issues of land and housing markets. In addition, a critical dialog between research contributions from the Global North and the Global South to the field of urban studies will be developed. The perspective on the role of planning will provide another theoretical input into this thesis. In the last part of this chapter, these multiple perspectives and their interrelations will be summarized.

2.1   Road Infrastructure Economics

The research on road infrastructure with respect to economic studies of the relation between highway construction and urban development is very relevant to this thesis. Mostly based on econometric analyses, researchers in this field have been plagued with the 'chicken-or-egg' question: Does new highway construction trigger new urban development or does new urban development lead to or necessitate new highway construction? This question exemplifies the common issue of figuring out causalities, while the research objects and the relevant parameters are highly difficult to define and assess precisely. Thus Sanchez concludes:

> "The literature on the effect of transportation infrastructure on the development of land is large but reaches few definitive conclusions and provides little empirical guidance." (2000: 3)

Some scholars have shown that highway construction decreases travel costs and makes far-out land more accessible, which decreases the price of developing this land, thus resulting in lower densities – the classical framework for an understanding of sprawl (for an overview, see Boarnet 1998; Sanchez 2000). Conversely, researchers point out that there are more variables to this logic than just the provision of roads (such as infrastructure and basic services or land development control through zoning). Furthermore, rather simple mono-centric forms of urban areas used in the applied econometric models do not represent the polycentric forms of many urban agglomerations (Boarnet & Haughwout 2000: 4). These objections are further supported by studies that show how highway construction in metropolitan regions has actually increased the price of land and led to higher densities along transport corridors (Sanchez & Moore 2000: 4; Boarnet & Chalermpong 2001: 576).

It all gets more inconclusive when inter- and intra-regional effects of transportation projects are considered. Scholars found out that urban growth induced by highway construction in one place might have happened anyway – just in another area of the metropolitan region (Sanchez 2000: 19; Boarnet 1998). While all these research contributions aim at figuring out the effect of building roads, the effect of *not* building roads is very much unknown and likely goes beyond the ability of econometric modeling for real-life examples (Sanchez & Moore 2000: 3). This might be even more relevant when different hierarchies of roads are considered and highway construction is not seen in isolation from road networks.

It has to be stressed that the literature on the impact of transportation infrastructure on urban development is very much centered on the United States and thus embedded in a particular context and history of both politics and planning. Nevertheless, the past years have seen some relevant contributions to road infrastructure economics with examples from the Global South, Africa in particular. Often directly interlinked with development agendas and/or programs of multilateral development banks and international agencies, there are two opposing research factions: one that seems to have a pro-growth bias and one that is rather skeptical with respect to the causality or the interrelation between road construction and (urban/rural) socio-economic development.

In what can be described as the predominant or mainstream literature on this topic, the role of road construction is often clearly linked to issues of tapping the underused economic potential of regions (Arnold 2005; Ndulu 2006; Calderon & Serven 2008). Road infrastructure is seen through a macro-economic and macro-regional perspective under the framework of developing functioning transport corridors that connect African countries and thereby ensuring the cost-effective and safe movement of goods and people (cf. Buys et al. 2006; Nathan Associates 2011; SOFRECO 2012). The 'underdevelopment' of these countries, it is argued (Ndulu 2006: 214), is rooted in dysfunctional, deteriorated, or non-existing basic infrastructures (particularly roads) and related public services (for instance, smooth customs controls at national borders through streamlined trade policies).

Some studies show the positive impacts of investing in infrastructure projects:

> "We find robust evidence that infrastructure development – as measured by an increased volume of infrastructure stocks and an improved quality of infrastructure services – has a positive impact on long-run growth and a negative impact on income inequality. The evidence also suggests that these impacts are not different in Sub-Saharan Africa vis-à-vis other regions." (Calderon & Serven 2008: 29)

However, what becomes apparent in these research contributions is this two-step causality model in which investment in infrastructure supports the realization of economic potential of 'sleeping' regions that then triggers social development – a logic that can be found in the well-known cost-benefit analyses of many infrastructure appraisal reports by development banks (as it is apparent also in the African Development Fund's appraisal report for the THIP, ADF 2007; also see ADF 2009, 2010a, b). First come productivity, efficiency, and growth; then, issues of access to basic services, equality, and human development are addressed. While this logic is deeply entrenched in the research literature as well as the development (investment) work of international agencies, one can ask why investment in social development would not be able to help regional economies to evolve to full potential – thus applying the economic-to-social growth argument in reverse. One challenge here can be seen in the missing link and even analytical challenge of connecting macro-economic and supra-regional transport corridor analyses to micro-economic, smaller-scale considerations of social and spatial development (Ayogu 2007: 115-117; Thomas 2009: 4; for Asia, see for instance Banerjee et al. 2012).

This aspect is taken up by the skeptical research faction. Concerning methodological issues, Button emphasizes the challenge of necessary data for working with causality models which "is an imprecise art at the best of times, and when there are significant lags involved it moves into the abstract school" (2002: 2-3). Furthermore, the seemingly clear findings by the 'pro-growth' research faction appear much patchier if the existing studies are analyzed in-depth. Ayogu in his literature review thus concludes:

> "A review of this empirical research reveals that the productivity effects of public capital vary from negative to positive and from small to large, with causality working in either direction." (2007: 82)

Connecting this back to the above paragraph, it becomes clear that causalities are not clear and that other aspects have to be taken into account when the impacts of large transportation infrastructure projects are to be analyzed. Here, the findings on inter- and intra-regional interdependencies with regard to growth come into play, where some regions experience increased development following heavy infrastructure investments, while regional systems overall lose their stability in terms of a spatially balanced economic and social development (i.e. transport corridor development bias; Salmon 2011: 16-23).

Eventually, every large transportation infrastructure project – even if it is meant to foster inter-regional trade connectivity – needs to be grounded in a smaller-scale setting (communities), which raises questions of integrating these projects into other policy programs. One prime example for this aspect is the Maputo Development Corridor (MDC) – a transport corridor between Mozambique and South Africa that has been developed since the mid-1990s to facilitate and revive inter-regional trade (i.e. the flow of goods and people) with an intended trickle-down effect on regional development (Louis Berger Group 2005: 9-17). Developed under the framework of the South African Spatial Development Initiative, this corridor is seen as having been designed under a neo-liberal paradigm that aims for less government and puts social issues aside in favor of global investment in an opening market, as Söderbaum and Taylor aptly describe:

> "Such strategies cast everything within a profit-seeking and 'bankable' framework which allows very little space for tackling the social and ecological implications that the various projects engender. (…) the MDC is designed first and foremost for 'big business', and local participation is more by coincidence rather than a guiding component (…) with the real intention to increase export growth and GDP [gross domestic product] rather than people-centred development. It is basically an 'investment initiative' of gigantic proportions (…)." (2001: 685)

Again, a causality is assumed in the logical framework of the Maputo Development Corridor: infrastructure investment supports economic growth that will trigger social development. It is not that authors like Söderbaum and Taylor (2001: 691-693) are denying the possibly positive impacts of such projects on the local scale. However, they are criticizing the narrow focus of infrastructure projects that are not integrated into the development of local capacities (also see Louis Berger Group 2005: 14-17). State functions are limited, commercialized, or privatized, with the state losing its mediating role to buffer global economic impacts on local communities. These generally promising typical public-private partnerships for large transportation infrastructure projects are not applied to their ideal functions:

> "What is emerging is not a partnership between state and capital in the service of the public good, but rather a deal between the political elite and transnational capital, supported by the IFIs [international financial institutions] and the donor community, to rush headlong into liberalisation." (Söderbaum & Taylor 2001: 687)

To summarize, the literature from road infrastructure economics reaches few definitive conclusions on the interrelation of infrastructure investment/construction, economic and social development, as well as spatial impacts on metropolitan areas or regions in general. A research field that has been heavily focused on case studies of the Global North, especially the United States, is increasingly turning to Global South examples. Therefore, one cannot yet expect extensive findings and a well-developed set of 'lessons learnt' – particularly in light of the rather young history of internationally-financed large transportation infrastructure projects in Sub-Saharan Africa.

11

Questions clearly arise from the objectives of such projects in consideration of targeted scales and underlying assumptions on causalities. Not only will the aspect of project impacts be discussed in this thesis, but also the relation between the state, the private sector, and affected communities. Various variables influencing the design, implementation, and outcomes of highway projects have to be taken into account. In this thesis, it will not be argued against the fact that the road infrastructure in many African countries is in a very bad state impinging upon daily lives in terms of both economic and social development (Kumar & Barrett 2008: x). Nevertheless, the mainstream 'trickle-down' argument needs to be discussed critically and the focus of road infrastructure economics has to be widened: What are transportation infrastructure projects for? Why is there so little consideration of non-motorized transport when transport networks are assessed? Why is the focus so much on road construction instead of road (traffic) management? These questions are directly related to the next research perspective of transport mobility and urban form.

2.2 Transport Mobility and Urban Form

As road infrastructure economics exemplifies, there is no simple formula for understanding how infrastructure provision impacts economic and social features in a particular setting. The real-life complexities of urban development, as it is happening in Sub-Saharan Africa, make it even harder to apply theoretical models. The literature that focuses on the interrelation between transportation infrastructure and urban form is rooted in the Global North and some of its concepts (such as smart growth) only seem partly helpful in exploring corresponding issues in countries of the Global South (cf. Angel 2011). However, one major conclusion from transport mobility studies is that the "freeway" is definitely not the right road design solution to deal with increasing traffic in urban areas (ITDP & EMBARQ 2012: 2). That hints at the conflict potential of the Thika Highway cutting through already developed settlements – and the following discussion of the interlinked concepts of mobility and accessibility will underscore this point.

While a broad presentation of mobility studies' findings from case studies goes beyond this thesis, UN-Habitat's Global Report on Human Settlements, titled "Planning and Design for Sustainable Urban Mobility", offers in-depth insights into the state of research and practice, underscoring the strong relationship between urban form and mobility. A major concern in this report is the challenge of sprawling metropolitan areas in the Global South – a process that is happening even more rapidly than it was experienced in the Global North (UN-Habitat 2013: Ch. 5). Even though case-specific contexts have to be taken into account, the UN-Habitat report's general statement summarizes the main problems:

> "The dispersal of growth from the urban centre is a worldwide phenomenon. Dispersal, as a form of decentralization, at least when it is poorly planned, lies at the heart of unfolding patterns of urban development that are environmentally, socially and economically unsustainable. With dispersal comes: lower densities, separation of land uses and urban activities, urban fragmentation, segregation by income and social class, consumption of precious resources such as farmland and open space and more car-dependent systems." (ibid.: 77)

In this setting, an important differentiation has to be made between the two main concepts of mobility and accessibility.[2] While mobility has been primed for many decades as the chief goal of improving transport systems, more recent studies identify issues of accessibility as being key for developing an urban system in which people can move, work, and live in a cost-efficient, socially equitable, economically prosperous, and environmentally sustainable way (UN-Habitat 2013: Ch. 5, Ch. 8; Beukes et al. 2011: 453; Cervero 2005). What distinguishes this perspective from the dominant reading in the road infrastructure literature is that contexts as well as political features are studied more closely. People have different needs and means. Institutions and political settings, power relations between stakeholders, provision of infrastructure beyond roads as well as social services all affect the functionality of an urban transport system that can support the achievement of other policy objectives. Therefore, context-sensitive design solutions are required for (large) infrastructure projects to further the achievement of broader policy goals.

While this thesis deliberately refuses to advocate for blueprint solutions that lack context-sensitivity, the transit-oriented development (TOD) model has proven to be a promising practice (UN-Habitat 2013: 93-94). TOD is rooted in Cervero's work on urban form and travel patterns with the famous 5 D's of density, diversity, design, destination accessibility, and distance to transit. This means that settlements are (re-) designed aiming at a compact form with short distances, thus being pedestrian-friendly. The urban form is oriented towards transit modes of mobility. Probably most important, the TOD paradigm aims at mixed settlements, where residential, commercial, educational, and leisure-related uses form 'complete' neighborhoods. While the actual implementation in an urban area poses multiple challenges in living up to these objectives, the laudable idea is the consideration of providing transportation infrastructure with consideration of concerned settlements and their spatial and socio-economic features.

Applied to a specific case, one can refer to studies on South African urban development. The setting in South Africa's urban areas is characterized by a Sub-Saharan typical stark segregation along socio-economic (and thus also often racial and ethnic) lines. Todes (2012a, b) analyzes the government's response to this challenging context in the case of Johannesburg's Spatial Development Framework (SDF). Early on, the SDF proved insufficient in tackling political, socio-economic, and environmental issues of the Johannesburg metropolitan region, because spatial planning was not linked with infrastructure planning. Due to the para-statal structure of infrastructure agencies, the state could not reach into all relevant policy fields in an orchestrated way. This has then partly changed with the development of a Growth Management Strategy that is strictly knotted to a Capital Investment Management System that strategically directs funds for infrastructure works in consideration of

---

[2] While very specific and complex definitions of mobility and accessibility exist, dictionary-based explanations of the two words are already precise: Mobility describes the ability of a person to move between places (Merriam-Webster Dictionary 2013b). Accessibility describes the ability of a person to enter and exit places (Merriam-Webster Dictionary 2013a). The former state is rather concerned with the flow between places and the modes of transport/moving, while the latter state is rather concerned with places (and related services etc.) being reachable/accessible and the conditionalities/variables of people to have this access. Both concepts are interdependent and go beyond physical/technical features, since other criteria such as social and economic in-/equality result in different abilities of people to enjoy mobility or accessibility. Several authors argue that accessibility is a more fine-grained challenge to achieve in an urban (transport) system (Cervero 2005: 1-4; UN-Habitat 2013: Ch. 5).

urban development objectives.[3] It goes without saying that the success of such an interlinked development approach requires long-term political commitment, sufficient financial and human resources, and a favorable implementation setting (Todes 2012b: 163). Nevertheless, this case highlights the importance of going from planning to actual implementation and benefitting from the interdependencies of transportation systems and urban development.

Literature on transport mobility and urban form shows how accessibility and mobility are two sides of one coin – they can be harmonized or are sometimes at conflict with each other (UN-Habitat 2009: 155-156). There is still a significant car bias in approaches to planning and managing transport systems that can lock in urban areas through a path dependency of misplaced investments (ibid.: 155-156; UN-Habitat 2013: 78-82). Even the concept of accessibility is not of help here if it is limited to an idea of access to private properties, resulting in a car-dependent, gated community-style urban fabric (Beukes et al. 2011: 453). Other elements, such as the role of public places, growth nodes, and mixed-uses of settlements, play a vital role. While questions of livability are often raised in the context of the Global North, major concerns in countries of the Global South still deal with the livelihood of urban dwellers.[4] This hints at the difficult task of translating lessons from one context into another (for instance from the United States to Kenya with regard to context-sensitive design or transit-oriented development).

The first two perspectives offered in this theoretical chapter illustrate how different views on large transportation infrastructure projects increase the understanding for their multiple facets and possible directions of impacts. Some interrelations of infrastructure development and settlement planning have already been pointed out; however, the THIP was implemented in the Northern Nairobi Metropolitan Region that is characterized by peri-urbanization with distinct features of land and housing markets. Therefore, a corresponding third research perspective needs to follow.

---

[3] A Capital Investment Management System is a GIS-based system in which capital investments by the public sector (in infrastructure and other program areas) are directly tested towards their linkage to the Growth Management Strategy of the metropolitan region and its development objectives. The budget allocation will then be based on a prioritization of public investment projects vis-à-vis their 'linkage' scores with the Growth Management Strategy (Todes 2012a: 406, 2012b: 163).

[4] With regard to definitions (cf. Hendriks 2010: 165-166), livelihood is understood in this thesis as the fundamental basis of an individual's life. This means that certain conditions are necessary to sustain the survival of that individual. These conditions concern ecological and health aspects (concerning harm against the body), economic aspects (with respect to occupations that provide the necessary means to afford things for daily life), social aspects (referring to the role of communities and social capital as well as educational features), infrastructural aspects (describing the physical-technical elements necessary to sustain life as well as their affordability and safety), and political aspects (meaning the political setting/context, in which an individual's life is not threatened or severely inhibited by means of abusive power). Some of these elements are reflected in the United Nations Millennium Development Goals. Overall, these aspects are deeply related to matters of accessibility.
On the other hand, livability – as it is understood in this thesis – builds up on a secured livelihood and refers to the quality of life and matters of improving the daily environment with regard to above-mentioned aspects. However, the focus is on an improvement of an already given 'quality', in that objectives of improvement are followed that are not solely based on survival necessities, but on broader political aims of making life more just, healthier, more efficient and more effective, more enjoyable, less resource-dependent, more inclusive, more democratic, more transparent, more in harmony with nature, more sustainable/long-lasting, etc. (cf. UN-Habitat 2009: 178). Therefore livability, as it is conceptualized here, can only be built upon a sufficient livelihood base.

2.3　Peri-Urbanization vis-à-vis Land and Housing Markets

Land and housing markets are of importance when studying Sub-Saharan areas with respect to urban development and related topics (such as infrastructure provision). This is even more so when peri-urban areas are considered. Before turning to the more specific literature on this region, a characterization of the peri-urban is necessary. Even though the following definition is derived from a European Union-funded research project on peri-urban development in Europe, it provides a detailed characterization that increases the understanding of a distinct spatial zone in urban agglomerations:

> "The peri-urban area is the dynamic transition zone between the denser urban core and the rural hinterland, consisting of a lower density discontinuous urban fabric and a mix of residential, commercial and leisure-related land uses. Peri-urban areas exhibit throughout Europe very different characteristics regarding spatial structure and density of the different land uses, ranging from continuous low density urban fabric, to scattered medium density settlements and commercial sites; from dense horticultural areas to arable and range land, to forests and natural areas. The peri-urban is not just an in-between fringe. It is instead a new and distinct kind of multifunctional territory, and often the location for opportunities such as airports, business parks and high value housing, which are all seen as essential to urban/regional development. However, in most cases, it is also the location for problems: urban sprawl, wasted public funds, traffic congestion, agricultural land under pressure, damage to landscapes and biodiversity, fragmented communities and social polarisation." (Piorr et al. 2011: 22)

Comparing this definition to other sources (cf. UN-Habitat 2005; GSAPP 2006: 1; Bengs 2005) as well as field observations in the Northern Nairobi Metropolitan Region (see chapter 3.5 introducing the study area), it becomes apparent that certain similarities in the characterization of peri-urban areas exist – which makes literature from both the Global North and the Global South useful for this analysis. There are shared issues; for instance, the conversion of land (most often from agricultural to residential/commercial), the lack of clear government responsibilities (regarding administrative boundaries as well as government scales), or the challenge of ecosystem services (in light of over-use and inter-regional imbalances).[5]

Scholars have come to the conclusion that the traditional ascription of urbanization phases to settlements is no longer sufficient, since a vast heterogeneity in the development of areas in urban agglomerations can be identified.[6] Thereby, peri-urban areas can exhibit lower as well as higher density development – again related to the problematic topic of causalities discussed above with respect to infrastructure provision and land development. Furthermore, peri-urban areas are not just functionally interlinked with the urban agglomeration. There is also a symbolic aspect

---

[5] See Annex A for a comprehensive illustration of driving forces and dynamics of peri-urbanization based on the EU-funded research project PLUREL (Piorr et al. 2011).

[6] This point is illustrated in a recent UNEP report: Traditional phases of rapid urbanization are followed by stabilizing urbanization and incremental densification, which increasingly mix with urbanization patterns of de-densification and re-densification, resulting in a large heterogeneity of such development patterns – the authors describe this as the "emergence of a lumpy 'rural-urban continuum'" (2013: 29).

of peri-urban areas being interlinked and thought of as part of a metropolitan region (Lambert 2011; GSAPP 2006: 5; Piorr et al. 2011: 24-25).

Besides these similarities, peri-urbanization shows some features in countries of the Global South, particularly Sub-Saharan Africa, that are distinctive (Kreibich & Olima 2002; Rakodi 1997d; UNEP 2013: 31). This is related to the rapid rural-urban migration. In a process of de-agrarianization[7] and increasing urban primacy[8], people flee from rural areas desperate for securing their livelihood and following an often disappointed dream of better life in the city (Adebayo 2005: 52-53; Oucho 2005: 89-90). However, cities in Sub-Saharan Africa are not fit to deal with the massive influx of people – Nairobi is no difference here (UN-Habitat 2010: 2). Urban areas are stretched beyond their limits concerning economic opportunities (i.e. livelihood-sustaining occupation), infrastructure and basic services, as well as social and ecosystem services[9]. In this process, cities are encroaching on their rural hinterland, thus transforming these areas into peri-urban transition zones that are fraught with conflicts in everyday lives with regard to infrastructure provision and government control of residential and commercial land uses (GSAPP 2006: 1; UN-Habitat 2005; Kante 2005: 246; McGregor et al. 2006).

In contrast to periods of rural bias or urban bias in Sub-Saharan Africa, when development programs by state authorities and/or international agencies were privileging one zone over the other, scholars as well as practitioners increasingly acknowledge that rural-urban dichotomies are neither helping to understand the rapid transformation processes nor are they supporting effective policies in dealing with above-mentioned challenges (Simon et al. 2006: 4-7; Parnell & Simon 2010: 56). It is still apparent that interlinkages between urban, peri-urban, and rural areas are not well understood. While case studies address these interdependencies, comparative work remains patchy. This might be due to the character of these studies. Mbiba and Huchzermeyer (2002: 115-118) criticize that much of the peri-urbanization research has been contracted by international development agencies, which resulted in an abundant body of descriptive studies. What is still missing so far are more critical analyses of peri-urbanization processes in which underlying causes and political aspects are discussed. The objective then would be to go beyond a characterization of peri-urban areas and to discover stakeholders and their interests in using/making

---

[7] The term de-agrarianization describes a long-term process in which rural populations adapt their economic activities to access new income sources, which is accompanied by changes to their social and spatial aspects of livelihood from more rural to more urban contexts (Oucho 2005: 90).

[8] Urban primacy is sometimes defined in quantitative terms, where the primate city in a country has a significantly larger population than the next largest cities and/or hosts a large number of the total population of a country. However, a more qualitative perspective on urban primacy is helpful to understand how the primate city's dominance in economic, political, and/or social terms can result in its congestion, the shortage of land, price increases, over-burdening of its ecosystem, and environmental degradation. At the same time, other urban areas receive much less attention and too little support in their development (UN-Habitat 2010b: 147).

[9] Concerning the terminology, Neuman defines infrastructure as "(...) built facilities and networks, which are above or below ground ('grey' infrastructure), and non-built, yet planned and managed landscapes that provide human and nature services ('green' infrastructure). This broad take includes publicly, privately, and jointly owned and operated systems (...)" (2009: 205). In these systems, he includes utilities, public works, community facilities, telecommunications, and green infrastructures. However, in this thesis, the term *infrastructure* refers to the actual concrete/physical material of utilities, public works, and telecommunications. The term *basic services* includes the works/activities that are necessary to provide, for instance, water, electricity, or waste disposal. Under the term *social services*, Neuman's "community facilities" are subsumed, including services in education, security, health, recreation, etc. Neuman's "green infrastructures" are termed in this thesis *ecosystem services*, including water bodies, arable land, forests, crops, etc. (ibid.: 205).

use of peri-urban areas in processes of rapid metropolitan growth. This is what this thesis will do when putting the THIP in its spatial context.

One crucial element in this regard is the discussion of peri-urbanization vis-à-vis land and housing markets. Edited volumes, such as by Rakodi (1997d), discuss the weakness of classical economic land and housing market theory in the Sub-Saharan context (and probably even beyond): The concept of a willing seller and a willing buyer that find each other on a perfectly efficient market is non-existing in real-life settings (Rakodi 1997c; Durand-Lasserve 2003: 3-4). There is a lack of information both in quantity and quality, the institutional framework is not sufficient, and access to 'the market' is not equitable, thus a blind faith in it must result in sub-optimal, i.e. often disastrous, outcomes (Rakodi 1997c). Well-known critiques of inefficient bureaucracy, missing or wrong incentives, too strict or not enforced regulations and codes, unclear legal systems, and lack of access to credit form the basis of such fragmented land and housing markets (ibid.). Formal and informal land and housing systems exist next to each other, are interrelated, or sometimes even amalgamated (ibid.). This impacts on a variety of actors, such as land and housing owners, renters, sellers, constructors, investors, buyers, squatters, etc. As a result, there is a highly insufficient provision of land and housing that qualifies as accessible, affordable, and safe (ibid.; Tacoli 2002: 10; UN-Habitat 2010: 154-159; Obala & Kimani-Mukindia 2002: 169).

This situation is aggravated with the fundamental role of land (and property on that land) in Sub-Saharan Africa. In some contexts, historically rooted cultural aspects have made land a valuable asset; in other settings, land has become an important asset through political and economic transformations (see for instance Cooper 2008; Klopp 2000). This is highly relevant to the topic of peri-urbanization, since land has a double nature as a "natural resource and marketable commodity" (Opoku Nyarko & Abu-Gyamfi 2012: 3). Land performs several purposes through its physical features, its cultural value, its socio-economic function, its political-historical role – there is hardly any policy area in Sub-Saharan Africa that is not related to issues of land (ibid.; Tacoli 2002: ii; Douglas 2006).

With the provision of infrastructure, access to and use of peri-urban land become more contentious. Particularly transportation infrastructure opens up peri-urban areas and thereby shifts the balance of what Calderon Cockburn (1999: 5-8) describes as the urban and rural functions of land profits: Increased access to peri-urban land through transportation infrastructure means a shortening of the distance to urban areas, increasing possible urban profits from that land (i.e. residential, commercial, industrial uses). This means that agricultural (rural) uses of that same land can only be preserved if rural profits from the land increase as well. Since government support for increased agricultural efficiency is often not provided, a process of land subdivision and conversion into urban uses is triggered. Farmers become 'speculators' by holding back their land from the market until seeking the 'once-in-a-lifetime' opportunity to sell (some of) their land for high profits (ibid.: 14; Tacoli 2005: 122). If speculation precedes the actual physical opening up of peri-urban or still rural areas, the development of land will happen in an uncoordinated, fragmented, and infrastructure-deficient way. Governments will have to follow with often under-budgeted reactions to this process of peri-urbanization (Calderon Cockburn 1999: 16). As a result, land is put into the highest individual uses instead of balancing it with "public welfare considerations" (Olima & Kreibich 2002: 6) – meaning ecosystem functions, equitable land access, incremental infrastructure provision, etc.

When Briggs and Mwamfupe describe this process for the Dar Es Salaam Metropolitan Region where peri-urban land develops from "a zone of survival" to "a zone for investment" (2000: 804), the authors point out another feature: If the political-economic setting in a country does not provide fertile ground for capital investment in long-term productive activities, people will seek the second-best option for securing their financial assets from being burnt in the next recession, inflation, or other (global or regional) economic-financial crises. It is in these situations that people turn to land. And multiple stakeholder groups will have different interests in putting peri-urban areas into agricultural, commercial, industrial, residential, or leisure-related uses (ibid.; cf. Douglas 2006).

What becomes clear in this part of the theoretical chapter is that the relation between peri-urbanization and land and housing markets is of key relevance to this thesis. Related to the other research perspectives presented, land and infrastructure are deeply interlinked. In processes of peri-urbanization, the provision of, access to, and use of land is even more accentuated. Specific challenges of managing urban growth emerge in these peri-urban zones. And these challenges are not only related to infrastructural or land features, but also to political aspects of governing metropolitan regions, which will be taken up in the next section.

2.4 Urban Studies in the Context of the Global South

This part of the theoretical chapter aims at applying research contributions of urban studies that are primarily concerned with case studies of industrialized countries to the context of urban areas in the Global South. While many scholars in their respective research field have not made their studies 'comparable' to different contexts, the reference to authors in the following sections will exemplify that this critical dialog is enriching for an understanding of the complexities in cities that are experiencing material and immaterial processes of globalization. Furthermore, it will form the foundation for a political perspective on public projects and their embeddedness in larger contexts of urban politics.

2.4.1 World/Global Cities and Entrepreneurial Urban Development

A broad-based excursion into world/global city literature is not necessary here, since many researchers have published extensive work on this topic (cf. for instance Brenner & Keil 2006; Sassen 2006; Taylor 2004). However, articles such as the one by Ancien (2011) provide a good overview of how world/global city literature emerged from a historical perspective onto the urban studies research agenda.[10] Ancien poses the question towards the various research contributions of how 'the global city' is actually (re-) produced. And this leads directly to her analytical framework of connecting world/global city literature with New Urban Politics literature. Here, research themes such as growth coalitions, public-private partnerships, or urban governance play a role. It is this thought framework of entrepreneurial urban development and achieving global competitiveness that forms the basis of many

---

[10] Ancien (2011: 2475-2477) presents this development as emerging from world cities literature by authors like Geddes, Hall, or Braudel to an urban studies perspective on 'world-cityness' by authors such as Friedmann and Wolff, Beaverstock, or Smith. This was followed by the 'global city' literature of Sassen or Fainstein, amongst others. Rather critical contributions by Massey or Smith discussed the 'global-ness' of places, while Marcuse and van Kempen looked closer into the 'globalizing' city and authors like Taylor focused on global city functions/networks.

cities and their politics – and which is of relevance to the analysis of the Nairobi Metropolitan Region and the THIP.

Studies by Loughborough University's Globalization and World Cities Research Network (GaWC, most prominently Taylor 2004) exemplify several points in this regard: Cities can be seen in a global network, in which connections between certain places make urban centers important nodes – in economic, financial, political, and other terms. Sub-Saharan cities are nearly absent in these analyses (Taylor 2004: 76-79). There is a missing link between the methodological assessment of global connectivity and the regional importance of cities such as Accra, Lagos, or Luanda. Besides this disconnect, cities in Sub-Saharan Africa – particularly the largest and/or capital cities – seek to participate in this global competition for being relevant nodes in an international network (Simon 1992: 48; Robinson 2006: 112-113; Lemanski 2007: 448-450; Grant 2009: 150-151). To be more concrete: It is urban elites in these cities that foster an entrepreneurial agenda for making their cities fit for this global competition disregarding fundamental needs of urban dwellers' livelihoods. As Simon in 1992 already pointedly described:

> "The more a square kilometre or two of downtown Dakar, Abidjan or Nairobi can be made to resemble Paris, London or New York, the less the elites tend to understand and tolerate the urban poor, except in so far as they represent an opportunity for exploitation and further capital accumulation." (155)

This understanding of 'global-cityness' is already useful, but other relevant urban politics topics that are rather disregarded in world/global city analyses need to be added. Most prominently, Robinson makes her point when calling for "post-colonial urban studies" (2006: xi) in which cities are seen as 'ordinary': The imposition of a Western modernist paradigm of how a city has to be, has to look like, and has to develop is criticized by her for often resulting in a detrimental socio-economic, political, or environmental development of cities in the Global South. Despite the fact that decision makers in these cities might cling to global city models of urban development, it is the contestation of these narrow thought frameworks that offers context-sensitive approaches to using the potential of urban dwellers to improve livelihood and create livability (ibid.: 110-114). Robinson correspondingly criticizes the status quo:

> "The politics of information circulation, international networking and conferences, advice and technical support, the prominence of donor agendas and consultants' analyses, as well as the effects of powerful discourses about urban growth and development: all these have the potential to limit local decision-making and impose external agendas." (ibid.: 127)

Lemanski in her paper on Cape Town (2007) has taken up this line of argument and discusses how these external standards and propositions to which urban elites subscribe channel scarce resources of cities in the Global South into a limited number of projects that put concerns of global competitiveness (for instance, concerning infrastructure and economic policies) above pro-poor policies or local needs in general. There are plenty of examples where cities in Sub-Saharan Africa display stark inequalities with an increasing spatial and socio-economic segregation, even though decades of neo-liberal market opening interlinked with global city agendas have proven to be detrimental to a majority of urban dwellers. In relation to that, scholars have challenged the idea that cities would be able to recoup the social, economic, or environmental costs of joining this global competition for becoming international nodes (Smart & Smart 2003: 270; Sassen 2006: 87) – the THIP has to be studied in consideration of these questions of needs and interests towards and costs and benefits of major public investments.

2.4.2 Urban Politics and Planning in City-Regions

Urban politics in light of world/global cities can only offer a comprehensive understanding if the research scope is widened: What makes an urban area experiencing globalizing transformations is not limited to the city itself, but extends into the urban agglomeration. This area has been a focus of the city-regions literature. While this part of the theoretical chapter does not seek to extensively discuss the Global North-centered debate on city-regions[11], it is argued that a city-region-wide view[12] provides further insight into the complexities of governing metropolitan areas, such as Nairobi.

Harrison talks about "multilevel metagovernance" (2007: 326) when trying to locate the city-region between local and national policies and institutional frameworks. The city-region becomes a conflict zone when decentralization and subsidiarity are advocated, but are not accompanied with the necessary fiscal devolution – even more so in Sub-Saharan Africa where many government systems are still highly centralized (Parnell & Simon 2010: 49). When going beyond (local) administrative boundaries questions of resources, representation, and responsibilities arise. There needs to be a discussion on how entrepreneurial urban development in the wake of global competitiveness affects the urban agglomeration: The 'urban hinterland' can become the 'dumpsite' of a thriving global city (cf. Etherington & Jones 2009: 260). Conversely, a socially just sustainability agenda can help to balance out negative spillovers into peri-urban areas if adjacent jurisdictions join forces (cf. Neumann & Hull 2009: 782-783). More recent contributions in this research field show that this happens increasingly in an informal style of government and governance (Allmendinger & Haughton 2009). Since the majority of literature is focused on the Global North and thus dominates the research agenda and debate on relevant topics, more in-depth studies of the governing of metropolitan regions in the Global South are urgently needed.

Related to that topic, Allmendinger and Haughton (2009) underscore in their paper on the large-scale spatial development and infrastructure regeneration project "Thames Gateway" in the London Metropolitan Region that these new semi-formal regional configurations in practice challenge theoretical models of public policy and planning. Silo thinking and stark hierarchies are partly replaced by experiments of making planning operational in a 'fuzzy' governance space. But before one jumps to the conclusion that this might be a new technical solution that can be applied in

---

[11] Cf. for instance Tewdwr-Jones & McNeill 2000; Ward & Jonas 2004; Harding 2007; Jonas & Ward 2007; Harrison 2007, 2010; Etherington & Jones 2009; Allmendinger & Haughton 2009; Jonas 2012.

[12] City-region is understood in this thesis as: "(...) a strategic and political level of administration and policy making, extending beyond the administrative boundaries of single urban local government authorities to include urban and/or semi-urban hinterlands. This definition includes a range of institutions and agencies representing local and regional governance that possess an interest in urban and/or economic development matters that, together, form a strategic level of policy making intended to formulate or implement policies on a broader metropolitan scale" (Tewdwr-Jones & McNeill 2000: 131).

metropolitan regions of Sub-Saharan Africa, the (re-) politicized nature of planning has to be highlighted here. This research perspective exemplifies that the city-region should neither be realized nor reified politically for its own sake (Jonas & Ward 2007: 170); however, it can offer approaches to balanced metropolitan development beyond formal, clear-cut administrative boundaries (for instance in the provision of transportation infrastructure). This is directly related to the design and implementation stages as well as outcomes of the THIP, since it was implemented in a functionally working, but administratively non-existing metropolitan region across local boundaries.

2.4.3 Spatial Segregation on Socio-Economic Grounds in Globalizing Cities

Following from a discussion of the world/global city literature and the view on politics and planning in the city-region, one further take on this topic complements the critical discussion of research contributions in urban studies applied in Global South contexts. It is a kind of micro-level view on how urban development is actually undertaken – in a setting where one perceives cities not as *global* but rather as *globalizing* (Grant 2009: 7, 13-15). This acknowledgement of an urban development diversity is also one reason why it makes little sense to analyze or seek to characterize 'the African city'. In this regard, scholars such as Simone (2004) have contributed to a rebuttal of analytically as well as practically questionable stereotypes of cities in Africa. Another example is Grant (2009), who paves the way for a more fine-grained reading of urban areas in Sub-Saharan Africa by identifying a hyper-differentiation in the case of Accra that goes beyond 'the African city', 'the world city', or 'the colonial city' – fragmented spaces where formal and informal, traditional and modern intermingle (ibid.: 7-8). While distinct frameworks of analysis such as the colonial or post-colonial city are not helpful, transformational elements such as the diversification of the urban fabric or the previously discussed processes of peri-urbanization seem more promising in comparing similarities and differences of Sub-Saharan cities.

In undertaking such an analysis of Sub-Saharan cities, matters of social exclusion are concerned in light of specific spatial configurations, in many cases related to spatial segregation on socio-economic grounds. With regard to the micro-level, disenfranchised groups are marginalized socially, politically, economically, and thus often spatially (UN-Habitat 2010: 2-3; Simon 1992. 147, Pieterse 2010a: 8). In these processes, the role of middle classes has gained momentum. What McGuirk and other scholars describe for case studies in the Global North (McGuirk 2007; McGuirk & Dowling 2011; Phelps & Wood 2011) can similarly be found in cities such as Nairobi, Abidjan, or Accra: Resembling Western modernist styles of middle class living (mostly influenced by the stereotypical image of middle class suburban living in the United States), materialism and clean, safe lifestyles characterize the new mantra of middle class living in globalizing cities in Sub-Saharan Africa (Mabin 2001; Pieterse 2010b: 13; Grant 2009: 42-63). Grant (2009: 42-43) sees this phenomenon as a reaction to the inundating informalization of land and housing. The 'stylish materialism' of privileged groups stands in stark contrast to average urban living conditions (ibid.: 63). Upper-middle income groups are not only attracted by Western lifestyle-marketed globalizing spaces, but also seek corresponding land and housing as an investment opportunity in an inequitable speculative market (ibid.: 63). As a result, these privileged market actors barricade themselves in master-planned estates (gated communities, private neighborhoods, etc.) and further contribute to spatial segregations in globalizing cities.

It has to be emphasized that these segregating processes can and do only happen in a certain political setting of disenfranchisement of less powerful and less wealthy citizens – in many cases the majority of urban dwellers (ibid.: 64-65; Linehan 2007). In this situation, public investments are also driven by the interests of a selected group of people in processes of urban politics that lack access to information and meaningful participation (Pieterse 2010b: 20). It sheds light on how large transportation infrastructure projects and public investments in general can be analyzed more comprehensively when one moves beyond apolitical, technical approaches of seemingly objective decision-making on basic and social services (cf. Rakodi 1997a). As Simon points out:

> "All too often, urban managers and planners still lack an appreciation of these underlying issues, having rather superficial perspectives or even operating in a supposedly atheoretical and 'objective' manner and thus basing their proposals primarily on (often inappropriate and decontextualized) technical criteria." (1992: 5)

From this results in many cases the 'fight' not against the roots of problems, but against those who have to suffer under them: it is a fight against the squatters not the dysfunctional land and housing markets, against the marginalized not the poverty, against the informal traders not the market conditions, etc. (Oucho 2005: 96). These ill-advised policies have a spatial dimension that is illustrated in works by Grant (for Accra; 2009), Pearce-Oroz (for Tegucigalpa; 2001), Mabin (for South Africa; 2001), or Linehan (for Nairobi; 2007).

Pearce-Oroz's case study of Tegucigalpa (2001), even though set in Central America, is particularly illustrative: Lower-income people in the capital city of Honduras had been living in very unsafe conditions in the inner-city due to limited access to land and affordable rental housing. Nevertheless, their chosen location provided access to the marketplace, forming the fundamental foundation of their livelihood. At the same time, a natural disaster like Hurricane Mitch in 1998 revealed their exposure to natural hazards as their informal settlements were destroyed. The government decided to resettle these people to an unpopulated rural area farther away from Tegucigalpa, without understanding that the availability of land was not enough in light of a much worse access to marketplaces and public services. An already existing socio-economic inequality developed into an even more detrimental spatial segregation.

In conclusion, this part of the theoretical chapter shows how world/global city literature provides insights into entrepreneurial aspects of urban politics in both, cities of the Global North and South. In relation to that, matters of participation, representation, and access (to political processes, land and housing, marketplaces, infrastructure and services, etc.) come into play. Which stakeholder groups influence urban development decisions? What interests are behind infrastructure investments and other urban development projects?

When cities develop visions for achieving 'global-cityness', the research focus has to be on disenfranchised groups. This also accounts for the corresponding implementation of urban development strategies, when questions of regional governance, seemingly technical, 'apolitical' planning, and pro-investment or pro-poor policies arise. A more fine-grained analysis of metropolitan regions is needed that goes beyond traditional, clear-cut categories and stereotypes of 'the African city'. What diversification can one find in the urban fabric of Sub-Saharan cities? How do globalizing impacts play out in the spatial patterns of these cities? How is the relation, or segregation, of the rising middle classes and the marginalized groups characterized?

While this part of the theoretical chapter has shown how urban politics is an important element in understanding the development of urban areas in the Global South, there is more to learn from a critical discussion of urban planning's role in this regard.

2.5   The Role of Planning

Planning literature reveals a weakness in terms of connecting theory and practice. While there are plenty of studies on practical planning cases and also academic contributions on the theory of planning and specific planning schools (and many do exist; cf. Banister 2012: 2-6), it is often not clear how theoretical foundations of this discipline translate into planning practice. One connection exists between the various urban planning visions or utopias and their positive as well as negative impacts on how cities have been designed and developed (UN-Habitat 2009: Ch. 3). That these visions are hardly ever translated directly into practice has to do with planning's position between the public welfare and market activities: The public sector most often cannot perform all implementation functions, but relies on the private sector to undertake some tasks (UN-Habitat 2012b: 109).

Planning – by definition – is more about anticipating and guiding than implementing. At the same time, the notion of a rational public planning would be misplaced. Even though some scholars suggest a simple technical role and objective function of planning, it has always been deeply political (ibid.). Therefore, it is of key importance to locate planning as one can find it in cities of the Global South in a political-historical context (cf. Rakodi 1997a). Otherwise, one would reach only a superficial understanding of how and why infrastructure projects like the THIP are designed and implemented as they are.

In its 2012/2013 State of the World's Cities Report, UN-Habitat provides this perspective in an overview (108-113; also see UN-Habitat 2009: Ch. 3): Planning systems in countries of the Global South were implanted by colonial powers – particularly with regard to French and British planning systems. Eurocentric planning models have influenced how the urban fabric in former colonial cities emerged. This means that master plans have been drafted, neighborhoods were assigned singular functions, cities were segregated along ethnic and racial lines, stark regulation and zoning regimes were introduced, technocratic approaches left no space for participatory elements in planning, and the overall system became inflexible and narrow-mindedly focused on formal and legal elements of the urban (UN-Habitat 2009: 57-59). After countries in the Global South declared independence from their colonial powers, international agencies and companies provided expertise to the new nation-states; however, it was again Western consultants that developed planning 'solutions' to 'problems' of 'the African city' and beyond:

> "From Asia to Africa to Latin America, 'master', 'blueprint' and layout plans have had similar, harmful consequences in countless numbers of cities: spatial segregation, social exclusion, excessive mobility needs and consumption of energy, together with poor regard for the

potential economies of scale and agglomeration that any city can offer." (UN-Habitat 2012b: 109)

In this UN-Habitat report, the authors frame this type of planning the "'Global Standard Urbanization Model of the 20th Century'" (GS20C; ibid.: 109) standing for a model that "privileges individualism, consumerism, new (artificial) values and lifestyles, excessive mobility and privatization of the public space" (ibid.: 109). As a result:

"(…) the GS20C model appears to be predominant across the world, being largely driven by land speculation and real estate interests that build cities according to financial and economic parameters often radically at odds with shared prosperity." (ibid.: 109)

Since many urban planners in the Global South have received training by Western institutions, the inheritance of the GS20C has been secured even when external consultancy is reduced due to various internal and external reasons (cf. UN-Habitat 2009: 52; Rakodi 1997a). In this setting, one can find an interesting system of *dys*functional planning in former colonial cities that nevertheless strongly performs the *function* of a powerful political instrument:

"It has been suggested that it may not always be in the interests of governments to reform their planning systems, as modernist planning places a great deal of power in the hands of government officials and politicians who might be reluctant to give this up. Modernist approaches are often land dependent, and authorities in many developing countries would not be willing to give up their control over land-related matters, as this would seriously weaken their position. Planning can be used as a 'tactic of marginalization', where particular ethnic or income groups are denied access to planning services and are then marginalized or stigmatized because they live in informal or unregulated areas. Another scenario is that urban areas are covered by rigid and outdated planning regulations that are only partially or intermittently enforced, and this opens the door to bribery and corruption. Master planning has been used (opportunistically) across the globe as a justification for evictions and land grabs." (UN-Habitat 2009: 58)

These descriptions are relevant when matters of informality and the regulation or exploitation of it are concerned – as the following chapters on the setting in Kenya and Nairobi, in particular, will exemplify. El-Shakhs (1997) correctly outlines that even if these inherited colonial planning systems were fit for properly managing cities in the Global South, they could still not function due to limited human capacities or capabilities, as well as financial resources of local authorities. Implementation is not a strength of lower-level government in Sub-Saharan Africa, especially given the often highly centralized government structure (Parnell & Simon 2010: 49; UN-Habitat 2010: 3).

Attempts of political decentralization have seldom been accompanied with fiscal decentralization or resource devolution, and this affects how larger-scale project plans, such as the THIP, are translated or interlinked with smaller-scale planning and development (cf. Wekwete 1997; UN-Habitat 2010: 3). In this regard, particular attention has to be on infrastructure, since it:

"(…) possesses a number of characteristics that endow it with leadership qualities. (…), without infrastructure, development does not happen. Infrastructure gets built first, shapes growth most, and lasts longest. Infrastructure links, enables, enriches, and provides opportunities and access." (Neuman 2009: 206)

As this quote illustrates, infrastructure – if planned comprehensively instead of in a piecemeal fashion – can play a vital role in addressing not only issues of physical improvements of the urban fabric. It also relates to political objectives in the field of social justice, economic growth, environmental conservation, public safety, etc. Again, it has to be seen vis-à-vis political contexts, stakeholders and their interests, and political-historical settings that have shaped urban planning and management systems in the Global South. Therefore, this part of the theoretical chapter complements research perspectives on urban politics and public investments by making the connection between the infrastructure perspective and the planning perspective under a political analysis framework.

2.6   Summary of Theoretical Perspectives

As illustrated in figure 2.1, this chapter brings together five theoretical perspectives on the study of large transportation infrastructure projects. These perspectives combine different research fields and theoretical approaches and together they form the basis from which the THIP will be analyzed with respect to some of the critical themes that were identified in the respective research fields.

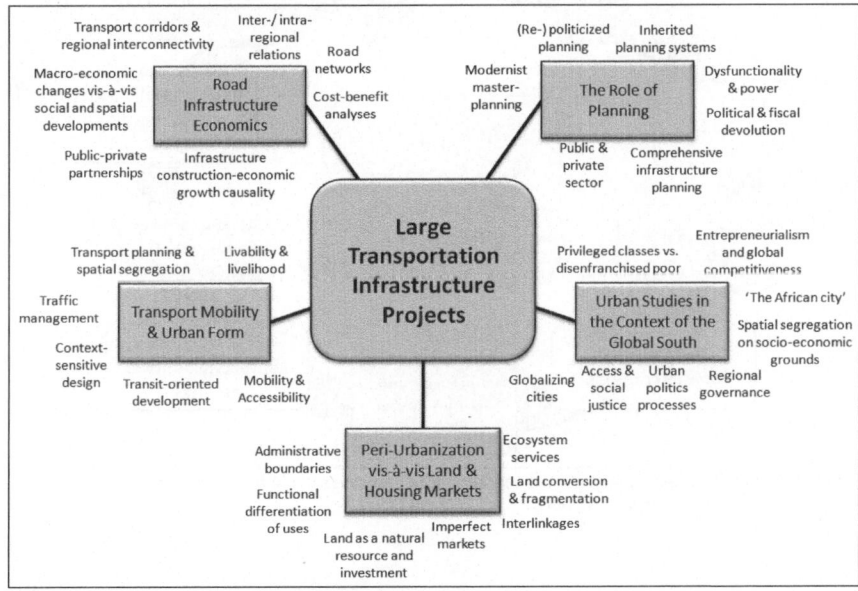

Figure 2.1: Five theoretical perspectives on large transportation infrastructure projects, with some of their exemplary critical themes (Source: Author).

In figure 2.2, the interrelationship between these five theoretical perspectives is outlined again. Starting from the road infrastructure economics literature, this chapter aimed at looking into the economic-engineering reasoning for large transportation infrastructure projects in relation to spatial and socio-economic development. Literature on transport mobility and urban form puts this perspective in a spatial context. Specifically related to the case study of this master thesis, the perspective from peri-urbanization literature with regard to land and housing markets helps seeing the interrelation between mobility and accessibility in a distinctive zone that calls for specific policy and planning approaches in the wake of rapid population growth and spatial expansion in metropolitan regions, particularly in Sub-Saharan Africa. In addition, this spatial context is juxtaposed with a political perspective in that urban studies literature can be fruitfully applied to urban development settings in the Global South. The heterogeneity of urban areas, as well as urbanization and globalization processes in the Global South is pointed out and the political variable in each of the before-discussed perspectives is accentuated. The fifth perspective derived from planning literature eventually helps to tie up rather technical and/or economic approaches with rather spatial and/or political approaches. As the (re-)politicized view on the role of planning already hints at, a more detailed political-historical background on politics and planning in Kenya and Nairobi, in particular, needs to follow.

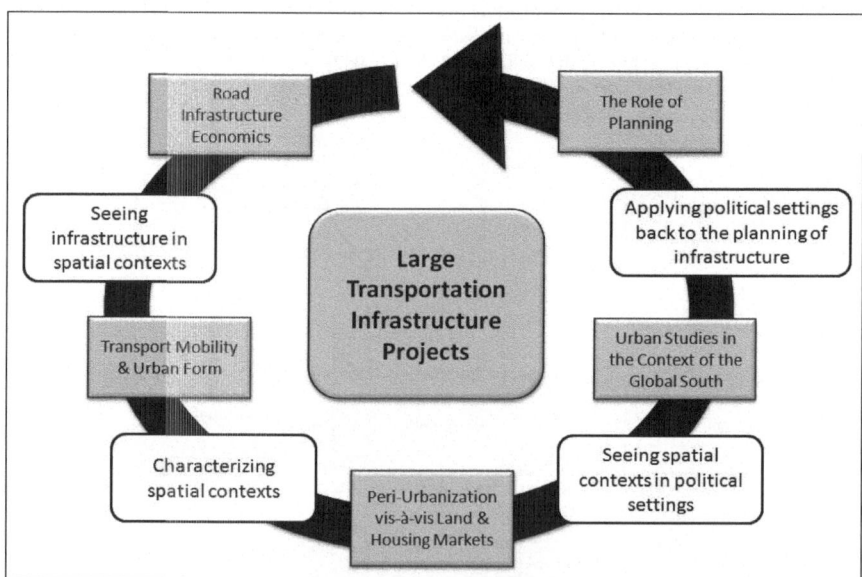

Figure 2.2: The interrelationship of five theoretical perspectives for the study of large transportation infrastructure projects (Source: Author).

3. Case Study Background

> *"(…) in Nairobi's history the language of planning has been used to mask key problems of spatial segregation and economic inequality." (Klopp 2012b: 5)*
>
> *"The net result is a city divided into distinct zones, from the ones who have everything to the ones who have nothing, and those in between." (Nabutola 2009: 2)*

Photo 3.1: Scattered development in the hinterland between Juja and Thika (Source: Author).

As the previous section on the theoretical perspectives on the research topic has shown, a critical introduction to the specific case study THIP and its context in Kenya is necessary. Thus, the aim of this chapter is to frame the context in order to enable a more comprehensive understanding of the THIP's idea, design, implementation, and outcomes in the empirical analysis. This is done by a political-historical introduction to the Kenyan state, followed by a discussion of urbanization challenges in Nairobi. Then, Kenya Vision 2030 and Nairobi Metro 2030 are presented and background information on the THIP and its context are provided. The last part of this chapter introduces the peri-urban study area in the Northern Nairobi Metropolitan Region.

3.1 A Political-Historical Introduction to the Kenyan State

There have been heated debates in academia about 'the African state'. While scholars could arrive at very precise descriptions of individual states and their history and development, comparisons and generalizations on the characteristics of the 'African' state have often been a difficult task. In his 1997 books review, Englebert discusses four contributions by Bayart, Dia, Mamdani, and Reno, which are aiming at a political theoretical characterization of the contemporary African state. While there are many labels for the African state as being a 'quasi-state', 'suspended', 'bifurcated', 'collapsed', or 'imported', etc., Englebert concludes:

> "The contemporary state in sub-Saharan Africa is not African. It descends from arbitrary colonial administrative units designed as instruments of domination, oppression and exploitation. (...) Nor is the African state a state. (...) In fact, it is because it is not African that the African state is not a state." (1997: 767)

If one, nevertheless, wants to characterize the state in Sub-Saharan Africa, the concept of the 'gatekeeper state', as described by Cooper (2008), might be revealing: The gatekeeper state can be ascribed to both the colonial and the post-colonial state system in Sub-Saharan Africa, since the basic conditions for this state to exist and to survive have changed only on their surface. Colonial rulers and as well as their post-colonial, independence-following successors have not been able to exercise their powers over the whole territory of the state. In addition to this territorial notion of a state, the notion of 'one' nation has also been questionable, since those in power have never been able to appeal to the huge diversity of different tribes or social and cultural groups in their respective country (ibid.: 5). Therefore, authors like Cooper see the label of the gatekeeper state suited to describe a state system, in which key decision makers' primary function is to "keep the gate" (ibid.: 157) in their individual interest of ensuring the smooth flow of goods and people (i.e. possible income sources) into and outside of the state.

It is telling that post-independence regimes on the African continent developed all kinds of political systems ranging from strong socialist to textbook capitalist ideologies (ibid.: 103), but the gatekeeper state stayed: Both colonial rulers and post-colonial 'liberators' feared liberal democratic systems, in which power would be derived from electoral legitimization for delivering on promises of shared growth. Instead patrimonialism was kept alive and, some argue, even resurrected in that ethnicity gained a political momentum – a segregated state along tribal lines was created to secure the ruling elite's power (Rakodi 1997a). This feature illustrates that the state system one can find today in many African countries is neither distinctively colonial nor post-colonial, neither traditional nor modern – it is an amalgamation and its very own thing which emerged out of European and African characteristics of ruling states and societies (Cooper 2008: 160).

What might appear to be far away from studying the THIP is actually relevant in understanding the framework of politics in Kenya (cf. Olima 2001; Kingoriah 2002; UN-Habitat 2012a). Focusing on the single most salient political issue of land, the Kenyan people look back on a history of spatial injustice and impunity with respect to land uses (Klopp 2000). It will be specifically related to Nairobi in the next section, but one can already see here the 'path-dependency' of segregation based on land emerging during the British colonial rule: Land management was the key tool to organize state, politics, and power (Olima 2001: 10). Once the colonists were ousted from power, the next group of powerful interests took over and quickly understood that the power over the land would ensure the power over the people (Kingoriah 2002: 216). What was later to be known as 'land grabbing' became a key characteristic of Kenya's ruling elite (Klopp 2000) – emerging under Kenya's first president, Jomo Kenyatta (1964-1978), and exacerbating under the regime of the second president, Daniel arap Moi (1978-2002). The entanglements between public officials and private actors (if they did not happen to be the exact same persons) in the land grabbing created long-lasting patronage networks to a degree that even the publicity about injustices in massive land allocations from the state to private individuals did not cause a transformative revolt by the people (ibid.; Simon 1992: 138; Kingoriah 2002: 216-217).

It is important to understand the instrumentalization of planning during these years. Excessive laws and regulations or their very absence; lack of implementation and enforcement; impunity in illegal land acquisitions and uses – they all characterized how land management and planning has been done in Kenya for most of its modern history (K'Akumu & Olima 2007; Olima 2001). But as the above-mentioned amalgamation of European and African features meant to underscore: It is not post-colonial decision-makers that are solely to be blamed. The post-independence state inherited an overly complex state system based on British rules and standards. Once it was removed from its original European (or colonial rule) context, the results were predictable to some extent (El-Shakhs 1997). What has been discussed in the theoretical chapter in terms of dysfunctional planning as a political instrument is further illustrated for Kenya by Kingoriah's description:

> "After 1980 or thereabouts, the senior officers at the 'Ministry of Lands and Settlement' and those of local authorities countrywide realised the extent of their immense powers over land-use planning, land-use, and land allocation. The social mood of corruption set in, both at local and central government level. The existing law that was based on British expectation of rational behaviour of 'ladies and gentlemen' fell apart in the presence of new crooks who were out to enrich themselves by acquiring as much real property as possible, especially in urban areas." (2002: 216)

In this situation, the formal system was increasingly bypassed, speculation in land increased, and the access to publicly available information on land became scarce (a feature that will be important again in the empirical analysis):

> "It is inconceivable for one to expect to be formally allocated a plot just because they had applied for one. The centralised administrative land allocation system has therefore suffered from malpractices such as rampant bribery, forgery, nepotism, favouritism and corruption (...). The system is being subverted and undermined by the same government officials who are expected to be responsible for its exercise. Thus, the practice has tended to reward the rich with no sensitivity to the urban poor, availing favourable access to public urban land for those close to the holders of power, or individuals and groups that win the favour of those in high office." (ibid.: 216)

In conclusion, this means that urban development, management, and planning challenges identified in modern Kenya and its urban areas cannot be understood comprehensively if this political-historical setting with its lasting impact on the Kenyan state and politics is not observed closely. And while the grabbing of public land reached its limits, this has not decreased the importance of land, since the few remaining public parcels increased in value in relation to supply (Klopp 2000: 17). Furthermore, extensive private land parcels (that in many cases are formerly illegally grabbed public land) become more relevant in land speculation. This aspect explains why under the third president, Mwai Kibaki (2002-2013), and even the latest president, Uhuru Kenyatta (since 2013), the fundamental role of land in Kenyan politics has not vanished.

3.2 Nairobi: A Globalizing City and Its Challenges

"Kenya's rapid urban growth comes with a host of problems, including high unemployment, the rapid proliferation of slums, overstretched and deteriorating infrastructure and services, choking traffic congestion, environmental disasters and the fast disappearance of urban greenery, an acute shortage of affordable housing and residential land amid financial speculation, high insecurity and crime rates (even though there are different views on the state of security), lacklustre or arbitrary law enforcement, petty politics coupled with political high-handedness, and all-pervasive corruption." (UN-Habitat 2012a: 2)

What is described in the above quote as urban challenges in Kenya precisely depicts the situation in Nairobi, as the following section will explain. Nairobi became the national capital in 1907, reached the city status in 1950 and has thus a relatively young history similar to several other African metropolises. Despite this young history, Nairobi has quickly become the dominant urban center in Kenya and can even be seen as the most important city in East Africa (Taylor 2004: 76-79; Simon 1992: 91-92; Obudho 1997). This role can be explained by the concentration of political and economic power that concentrate in the metropolitan region. In addition, Nairobi has attracted people from inside and outside the country, thus having been growing from some 200,000 residents in the 1950s to over 3 million in roughly half a century, while sustaining an annual growth rate of four per cent (UN-Habitat 2010: 8, 244). There are no signs that this urban primacy of Nairobi will be challenged in the short or medium term, thus the relevance of its peri-urban areas, where most of the growth happens, is likely to increase further (Otiso 2005).

One perspective on Nairobi as an East African hub could be the macro-level view of the GaWC on global city networks (critically discussed in chapter 2.4.1; cf. Taylor 2004: 76-79). With the United Nations system in Nairobi (UNEP and UN-Habitat headquarters and more than 20 presences of other UN institutions), the city attracts personnel and their families of international agencies as well as of the foreign arms of countries (embassies, chambers of commerce, cultural centers, etc.). Nairobi has also developed into one of the most important places for non-governmental organizations (Taylor 2004: 99). In addition to these political features, the economic diversity of Nairobi and its striving for knowledge industries has attracted much attention at least for market actors in the Global South.

This characterization tells something about the context in which urban issues have been influenced during the past two decades at least. Infrastructure provision, housing construction, urban design, and land use zoning have all been catering to the needs of an aspiring "regional hub and a potential 'global city'" (Hendriks 2010b: 136). Simon already identified in 1992 (97) that related land and housing investments can be interpreted or factually are speculative investments that aim for short-term profits, while money from long-term investments, such as in the production sector, is withheld (cf. Briggs & Mwamfupe 2000: 804).

The combination of Nairobi's spatial extension, its population explosion, and its 'global city' urban development focus (Linehan 2007: 22, 25) has resulted in abundant challenges. As it is summarized in various sources (GOK 2008b: 17, 42; UN-Habitat 2012a: 2, 8; Olima 2002: 59-60), Nairobi and the whole metropolitan region experience immense pressures on the ecosystem and corresponding resources, such as water and open green land. Land use planning and management has been non-functional for decades, resulting in an urban system that is neither informed nor controlled by responsible authorities (Olima 2002: 55-61). Prices for land and housing have been exploding. Informal settlements and slums[13] have been proliferating (Otiso 2002). Congestion and pollution in the city have been pressing concerns (GOK 2008b: 17). The municipal authorities are described time-and-again as ineffective in dealing with these challenges – with regard to financial resources, human capacities and capabilities, institutional structures, bureaucracy, as well as matters of coordination and political will (ibid.: 17; Office of the Prime Minister & MoLG 2010: 11-12).

Descriptions like these introduce the situation in the Nairobi Metropolitan Region; however, they do not dig into the underlying reasons nor do they provide a comprehensive picture of the challenges. Therefore three elements shall be further presented in this chapter: the issue of inequality, the characteristics of land and housing, and the state of transportation in the Nairobi Metropolitan Region. These elements are of particular importance when the THIP is analyzed in the empirical chapter of this thesis, as they help to frame the context in Nairobi, which is characterized by spatial segregation on socio-economic grounds, where matters of mobility and accessibility regarding settlements and transport are concerned.

3.2.1 Inequality

Not uncommon for Sub-Saharan Africa, Kenya displays a strong income inequality, with its capital city scoring a 0.59 Gini coefficient[14] (UN-Habitat 2010: 142). This inequality in Nairobi cannot only be grasped by quantitative data but also by qualitative assessments (Nabutola 2009: 2; Hendriks 2010b: 139-140). As has already been explained, the spatial segregation that characterizes Nairobi has its roots in both colonial and post-colonial policies of the gatekeeper state (Olima 2001). Nevertheless, an important change can be observed in the more recent development of Nairobi: In contrast to more rural areas in Kenya, segregation in Nairobi has been

---

[13] There is a controversial debate in both research and practice about defining informal settlements and slums. More recently, both terms have been used increasingly together, even interchangeably. The author of this thesis opposes this imprecise working with two different types of settlements. Even though UN-Habitat itself has increasingly used both terms together, their earlier slum household definition is clear to work with: "A slum household is defined as a group of individuals living under the same roof lacking one or more of the following conditions: access to improved water; access to improved sanitation facilities; sufficient living area (not more than three people sharing the same room); structural quality and durability of dwellings; and security of tenure" (2008: 92). This definition means that a slum overall combines households sharing these deficiencies to a varying extent. In contrast, an informal settlement is primarily characterized by having been put up illegally and/or being placed illegally in a certain location. While the definition of 'illegality' is context-dependent and heavily impacted by political features on the ground, it helps to provide an understanding of a settlement that is at risk of being evicted by government. Besides this feature, an informal settlement can exhibit the same or even better standards than formal settlements. One has to acknowledge the large heterogeneity (in-) between slums as well as informal settlements (Hendriks 2010b: 58), but they should not be confused as the livelihood of urban dwellers in slums is severely put at risk.

[14] The Gini coefficient is a commonly used statistical measure of income inequality, with figures closer to 0 meaning a more equal income distribution and figures closer to 1 meaning a more unequal income distribution.

shifting from traditional ethnic lines to socio-economic criteria (K'Akumu & Olima 2007: 89-90). While some authors describe the city as being divided into three groups of 'the rich', 'the poor', and a somewhat fuzzy 'middle class' (Nabutola 2009: 2), other authors see a more distinct gap between the privileged better-off and the excluded 'dirt poor' (Linehan 2007: 35; K'Akumu & Olima 2007: 96). It is not productive to reach a definitive conclusion here. It is more revealing to point out that Nairobi has high income disparities which reflect on the expenditure distribution of its citizens. In relation to that stands an unemployment rate of approximately one fifth of the working population and two fifths amongst the youth, which poses extreme socio-economic challenges (Hendriks 2010b: 139). While the informal sector employs many people – some authors state three quarters of the workforce (ibid.: 100) – and ensures some kind of livelihood (Otiso 2002: 255), this thesis explicitly follows the argumentation by some authors who caution against misinterpreting the *functionality* of the informal system as a *solution* to socio-economic inequalities:

> "Informal employment is nothing but a survival strategy for those excluded from the formal labour markets, in the same manner as slums are the residential survival strategy for people excluded from inefficient formal land and housing markets." (UN-Habitat 2010: 145)

This quote shows the two sides of the 'informality coin' in that inequalities in cities like Nairobi can only be understood by considering where people live *and* work. Linehan (2007) goes one step further by situating this spatial dimension in a context of economic investment interests that have been overwriting demands for social justice in the globalizing city Nairobi. These challenges need to be seen through a spatial lens and the following quote therefore connects inequality with the issues of land and housing:

> "Nairobi is consequently a city of walls and boundaries, a city of enclaves, a fortress city – a model of urban development repeated throughout the continent and in many places throughout the Global South (...). Underwriting this urban topography is a social and political process, which maintains the exploitative system of connections and strategic disassociations that characterize the spatial form in the city." (ibid.: 35)

3.2.2 Land and Housing

Working on the issue of formal and informal land uses in the Nairobi Metropolitan Region and referring to corresponding statistics can cause protracted discussions. Nevertheless, researchers that have been studying this topic for the case of Nairobi seem to come to an agreement that roughly 60 per cent of Nairobi's urban population lives on only five per cent of the city's ground. Most of these settlements can be characterized as informal. And one third of the city's population is believed to live in slums[15] (Olima 2001: 11; UN-Habitat 2010: 153).

The problems of land and housing have been manifold. It is a question of access to land and housing, tenure security, the quality of land and housing, conflicting titles, excessive rates for both land and housing, lacking enforcement of regulations, but at the same time also too high standards for formal housing, etc. (Otiso 2002: 257; GSAPP 2006: 2, 13; UN-Habitat 2012a: 7; K'Akumu & Olima 2007; Olima 2002: 60;

---

[15] See footnote 13 for a differentiation between informal settlements and slums.

GOK 2004: 2). Land and housing development in Nairobi and Kenya in general is almost exclusively developer-driven (Hendriks 2008: 31; Nabutola 2009: 11). Land owners are attracted by short-term income and subdivide and sell their land when the chance arises. In this case, land uses rapidly change often without the planning authorities controlling or even noticing the changes, which is why, for instance, agricultural and open green land are often irreversibly lost at a tremendous pace (Knight Frank 2012: 3; Thuo 2010: 7; UN-Habitat 2010: 168).[16]

Lower-income citizens are forced to opt for informal settlements, most often on the urban fringe in peri-urban areas, where land and housing are still affordable to some extent (UN-Habitat 2010: 153). The empirical analysis will therefore discuss how the opening up of peri-urban areas through infrastructure provision, such as the THIP, impacts on land uses and affordability. In most cases, smaller-scale infrastructure is only provided for higher-income estates, particularly the rising number of gated communities, while informal settlements and the maintenance of existing infrastructure are disregarded (GSAPP 2006: 17; Olima 2002: 60). As a result, people are forced to turn to illegal uses of infrastructure and basic services. This over-use of resources in land, water, and electricity puts increasing pressure on the ecosystem and threatens the functionality of the Nairobi Metropolitan Region. In many regards, the metabolism of Kenya's capital city is already stretched beyond its limits (Obudho & Juma 2002: 44). In this context of state withdrawal and higher-income-driven demand, land and housing have experienced an extreme commercialization (Hendriks 2008: 27; cf. Mbiba & Huchzermeyer 2002: 121). At the same time, elements of citizen participation in planning and managing the scarce resources of the Nairobi Metropolitan Region have been absent (Linehan 2007).

When land and housing are seen in light of the previously introduced interlinked concepts of mobility and accessibility, a discussion of transportation in Nairobi reveals further insights into the interrelated growth management problems of the metropolitan region.

3.2.3 Transportation

Rooted in the original transportation and urban planning during the British colonial rule, Nairobi expanded in a star-shaped pattern following major transportation corridors (GSAPP 2006: 14; Becker 2011: 2).[17] While the Nairobi Metropolitan Growth Strategy of 1973 identified several projects for constructing and expanding roads, most of these have not (yet) been implemented (Becker 2011: 2). Transportation planning for the Nairobi Metropolitan Region is segmented along horizontal and vertical lines of ministries and para-statal agencies, as well as national and lower levels of government (KIPPRA 2005: 78). Furthermore, there is a lack of coordination by international donors, who are currently providing most of the resources for the city's and the country's large transportation infrastructure projects (Becker 2011: 6; Klopp 2012b: 13). This raises questions about how strategic investment and project decisions further larger growth management policies.

Applied to a specific example, such as the THIP, this institutional description still leaves open an important question for understanding the functionality of

---

[16] See Annex B for three maps prepared by Mundia and Aniya (2006), which clearly illustrate the rapid and vast land-use changes in the Nairobi Metropolitan Region over only a quarter of a century.
[17] See footnote 16.

transportation infrastructure in Nairobi: How do people actually travel? Salon and Aligula (2012) undertook the first comprehensive study on travel behavior in Nairobi. Similar to other metropolises in Sub-Saharan Africa, the majority of citizens, particularly lower-income groups, first of all use non-motorized transport – i.e. they walk[18] (ibid.: 69). Middle-income groups and to a lower extent lower-income groups depend on public transit (ibid.: 69), although how Salon and Aligula correctly point out: "Matatus constitute the paratransit system in the city; there is no formal transit system to speak of" (ibid.: 72). The minibus (matatu) system in Nairobi provides a good geographical coverage, but the quality is sub-standard by all other means of safety and service (ibid.: 72; Klopp 2012b: 6-8). Only higher-income groups depend on individual motorized transport via private cars (Salon & Aligula 2012: 69). The status symbol of private car ownership does, however, pose a tremendous challenge to the urban transport system if one considers the rise of the Kenyan middle class and their likely consumer behavior (GSAPP 2006: 28). Until then, it remains a conflict between the need for decent infrastructure for non-motorized transport and the current car-bias in transportation planning in Nairobi in a setting where transport costs already reap a large share of urban dwellers' household income (Salon & Aligula 2012: 71). As Salon and Aligula summarize:

> "Transport planners in Nairobi have the problem of substantial numbers of people needing basic mobility services at the same time that the roads are highly congested and the social and environmental externalities of their transport system are large. The fact that Nairobi is a rapidly growing city makes the transportation challenge significantly larger." (2012: 65)

In conclusion to this part on the city's challenges, it becomes clear that transportation, land and housing, as well as inequality deeply affect how the metropolitan region (not) functions and develops. Before these elements are taken up in the empirical analysis, the THIP has to be seen not only vis-à-vis its spatial context in Nairobi, but also with respect to its embeddedness in Kenya's development vision.

3.3   Kenya Vision 2030 and Nairobi Metro 2030

In order to understand current investment decisions in Kenya, amongst them the THIP, the broad visioning process in the country needs to be introduced. This process started under the presidency of Mwai Kibaki, the third president of the Republic of Kenya (2002-2013). While the framework document "Kenya Vision 2030" (GOK 2007) reveals much about how the decision-makers imagine the development path of the whole country, the strategic plan for Kenya's capital city "Nairobi Metro 2030" (GOK 2008b) concretizes the translation of Kenya Vision 2030 objectives into specific projects for the Nairobi Metropolitan Region.

3.3.1 Kenya Vision 2030

> "Kenya Vision 2030 is the country's new development blueprint covering the period 2008 to 2030. It aims to transform Kenya into a

---

[18] Bicycling is considered too dangerous by many transport users, bicycle lanes are quasi non-existing, and bicycling is furthermore not common for Kenyan women (Salon & Aligula 2012: 72; interview UN-Habitat expert A; World Bank 2002: 144).

newly industrialising, 'middle-income country providing a high quality life to all its citizens by the year 2030'." (GOK 2007: 1)

This introductory quote from the original document already sets the frame in which the vision can be analyzed. Based on an economic, a political, and a social pillar Kenya is envisioned to be developed – or even stronger: transformed – from a 'developing country' into an 'industrializing country' by 2030 through five-year medium-term plans that are successively implemented. As a foundation of this growth path, so called "enablers and macro" are listed (GOK 2011a), with infrastructure (particularly large flagship projects) being primed as a key basis for the implementation of other projects in the vision.[19] This hints at THIP's idea, which will be further scrutinized in the empirical analysis.

Besides the aspect of an "all-inclusive and participatory stakeholder consultative process" (GOK 2007: 1) through which the vision was (claimed to have been) developed, the document definitely convinces even critical readers by its blunt openness about Kenya's strength and weaknesses (ibid.: 6, 19). Most challenges that were outlined in previous sections of this thesis are also identified in Kenya Vision 2030 documents (see for instance GOK 2008a). Even though worthwhile objectives with regard to social conditions and democratic procedures are expressed (GOK 2007: 19, 25), the document is dominated by an entrepreneurial, world-city featured language of global competitiveness (ibid.: 13, 15).[20] One does not need to judge about this framing, although it is important to be kept in mind when aspects of the THIP and its outcomes are discussed at a later point.

3.3.2 Nairobi Metro 2030

With respect to Nairobi Metro 2030, there are different ways of looking at it: One could see it as a second-order visioning document following Kenya Vision 2030. Another perspective would be to see this document as a strategic plan to realize the objectives that were identified for urban areas in Kenya Vision 2030. Without any doubt, the two documents are interrelated. In the framing of Nairobi's vision a similar language as in Kenya Vision 2030 is applied: It is mainly about achieving global competitiveness, providing world-class infrastructure, and developing a service center for East Africa and the global market by playing the role of a strategic hub (GOK 2008b: iii-viii):

> "Nairobi Metro 2030 is the Nairobi Metropolitan Region's (NMRs) statement that it aims to grow and develop into a world class African region, that is able to create sustainable wealth and offer a high quality of life for its residents, the people of Kenya, investors and offer an unmatched experience for its esteemed visitors." (ibid.: iii)

For that, six core values are identified: "innovation, enterprise, sustainability, co-responsibility, self-help, and excellence" (ibid.: v). The document includes relevant urban development and livability features, although its authors seem to have been more attracted by creating an effective place branding similar to what their chosen

---

[19] The corresponding website is also informative as the overall vision, its pillars, and the projects with their background and implementation status are presented: http://www.vision2030.go.ke/ (GOK 2011a).
[20] At this point, one could further discuss the role of McKinsey's consulting input into Kenya Vision 2030 (Linehan 2007: 24, 28).

reference points in South Africa and Australia have done (ibid.: v). Again, the question arises of whose needs are actually catered for with such a vision and corresponding policies (Linehan 2007: 28, 36; Klopp 2012b: 10-13).

Chart 1-4: The Nairobi Metropolitan Region

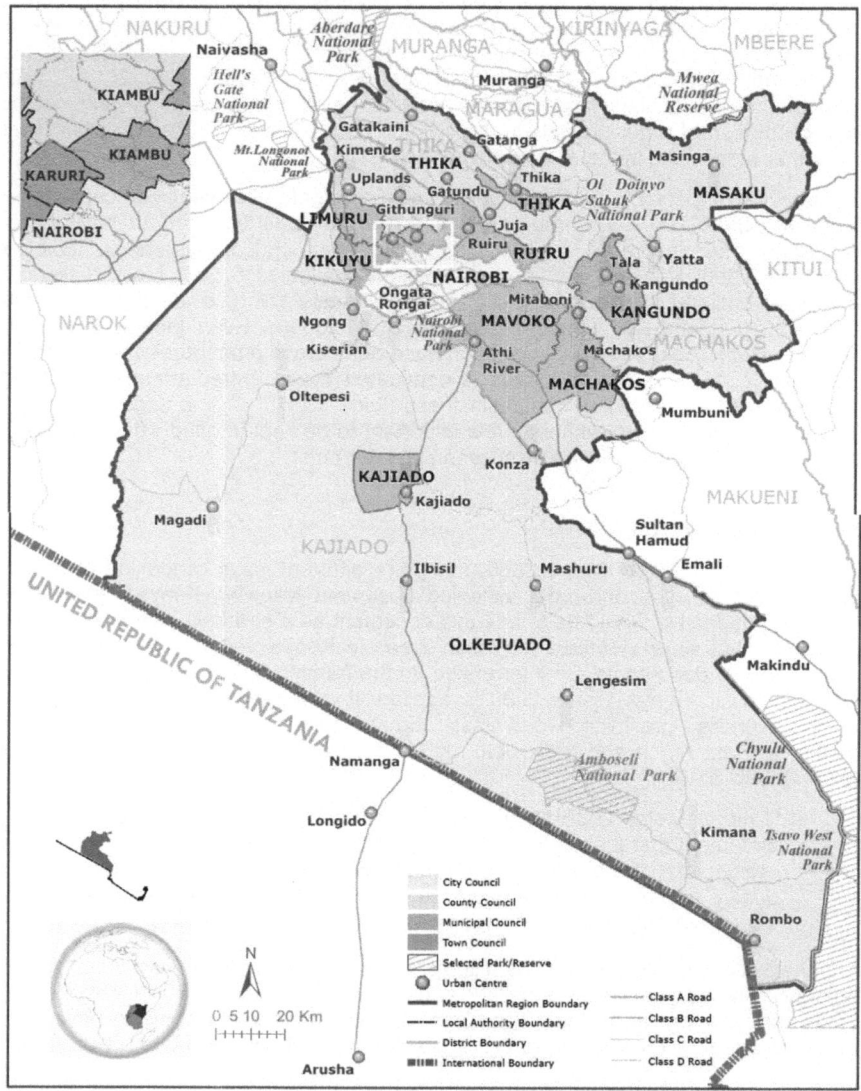

Figure 3.1: Map of the Nairobi Metropolitan Region as identified by the Kenyan Government (Source: GOK 2008b: 9)

Referring back to the theoretical discussion on city-regionalism and regional governance, Nairobi Metro 2030 (GOK 2008b: iv) is very telling in that the vision refers to an imagined metropolitan region going beyond Nairobi's administrative boundaries and including 15 local authorities (see figure 3.1). This is a clear political take on governing the Nairobi Metropolis. In contrast, the notion of the "Nairobi Metropolitan Region" as it is applied in this thesis stands for the functional and symbolic reach of Nairobi's impacts on adjacent urban, peri-urban, and rural areas (see figure 3.2).[21] In how far this corresponds to issues of regional governance will be discussed further.

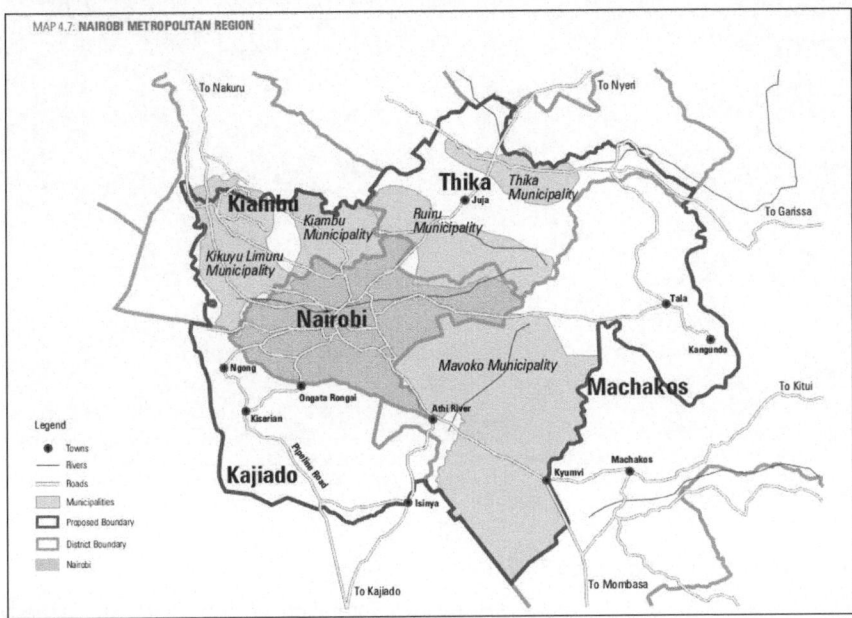

Figure 3.2: Map of the Nairobi Metropolitan Region as identified in research literature (Source: UN-Habitat 2010: 168).

Since both the vision for Kenya and the particular strategy for Nairobi are directly interconnected documents, in the following analysis this package will be referred to as "Kenya Vision 2030", always including Nairobi Metro 2030 – this was also a common reference and labeling in the conducted interviews for this thesis.

Besides these two visions, there exists a large amount of other plans for Nairobi and its metropolitan region regarding various topics such as land use planning and transportation planning (see for instance JICA 2006; Ortiz 2011a, b; Consulting Engineers & Runji & Partners 2011). While a discussion of these documents goes beyond this thesis, one can remark that the consultancies behind the drafting of these plans (often contracted by international agencies and donor countries) most often follow the textbook world-city visions despite the fact that they have been criticized in many different contexts all over the world – particularly because they have not proven to sufficiently address sustainability concerns with regard to

---

[21] See footnote 12 for a definition of city-regions related to this functional and symbolic understanding.

economic, social, environmental, or political issues (as has been argued with respect to planning in chapter 2.5).

Photo 3.2: Old Thika Road before the THIP (Source: Amboseli Daima, http://www.freeimagehosting. net/uploads/125762a9bd.jpg).

3.4   An Introduction to the THIP and Its Context

In the introduction of this thesis, the major information on the THIP was already provided. It is an approximately 50-kilometer long road that was expanded from a dual carriageway into an eight-lane highway from the end of 2008 until the end of 2012 for 31 billion Kenyan Shillings (approximately 370 million US-Dollars), financed by the African Development Fund of the African Development Bank Group, the Export-Import Bank of China, and the Government of Kenya. The construction of the three sections were undertaken by Shengli Engineering Construction Group Co. Ltd., Sinohydro Corporation Ltd., and China Wu Yi Company Ltd., respectively. This A2 international trunk road can be seen as one part of the to-be-realized Great North Trans-African Highway route going from Cape Town in South Africa up to Cairo in Egypt (see figure 3.3).

In order to understand the official reasoning for the THIP, the appraisal report by the African Development Fund (ADF 2007) can be referred to: This document describes the project as aiming at a reliable, affordable, and accessible transport infrastructure system that will foster socio-economic growth and development which will benefit the specific population in the Nairobi Metropolitan Region, Kenyans in general, and even neighboring countries (ibid.: vi). The project's main focus was defined in the appraisal report as releasing the heavy traffic congestion, decreasing the corresponding environmental and health impacts and excessive travel times, supporting a public

38

transit system, decreasing the high number of accidents on the road, and increasing the economic output through decreased travel costs (ibid.: vi-vii).

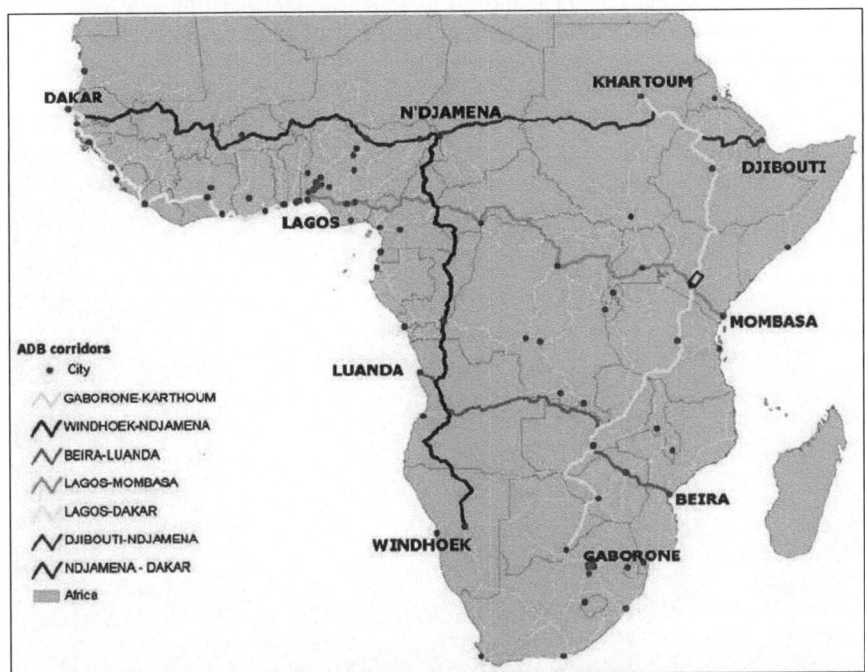

Figure 3.3: Map of the African Development Bank's proposed highway corridors in Sub-Saharan Africa; small blue rectangle in Kenya, East Africa, depicts the THIP as part of the Great North Trans-African Highway route, which is shown here as the Gaborone-Karthoum route (Source: Buys et al. 2006: 4; blue rectangle added by author).

Based on an environmental and social impact assessment, proper consultations and compensations for incurred losses/damages were to be done (ibid.: 11-12). The ADF formulated the objective of including international best practices into the project (ibid.: 8). Furthermore, special attention was given to address issues of pro-poor measures and gender equity (ibid.: 18). The whole project was put through a cost-benefit analysis in which benefits came out higher than expected costs (ibid.: 17-18).[22] In the empirical analysis, a critical discussion on the content of the appraisal report with regard to beneficiaries, expected costs and benefits, objectives and outcomes will be provided.

Bringing together the THIP, Kenya Vision 2030, and Nairobi Metro 2030, it is important to recognize the whole package of road infrastructure extension in the country. In the Nairobi Metropolitan Region, already completed or currently under construction are the Northern, Eastern, Southern, and the Greater Eastern Bypasses

---

[22] In case of a negative result, the project would have needed to be adapted significantly or would not have been supported by the African Development Fund – although, this would likely have not deterred the Chinese partners from financially supporting and implementing the project, if necessary funds could have been secured (cf. The Design of the THIP on the design of the THIP and the actors behind it).

(GOK 2011a). A relevant national road project in relation to the THIP is the LAPSSET Corridor Route (Lamu Port and Lamu-Southern Sudan-Ethiopia Transport Corridor)[23] that reaches from the port of Lamu up to South Sudan and Ethiopia over 1,730 kilometers crossing the Thika Highway at Isiolo (i.e. 250 kilometers beyond the end point of the THIP of what is already part of the Isiolo-Merille Road project) (GOK 2011a).[24]

Photo 3.3: Scattered densification development in Juja (Source: Author).

3.5   An Introduction to the Peri-Urban Areas along the THIP Corridor

The last element of the background to the case study is the characterization of the study area in the Northern Nairobi Metropolitan Region. Although it is not easy to find exact population data on settlements in Kenya, one recent report estimates the population of the Nairobi Metropolitan Region to be roughly 6.7 million inhabitants in 2009 (Consulting Engineering Services & Runji & Partners 2011: 6-7). Combining the data of areas adjacent to the THIP corridor (cf. figure 1.1) an estimated number of around one million people lives in these settlements that had been part of the City of Nairobi, the Municipalities of Ruiru and Thika, as well as Thika County Council.[25]

---

[23] The background and progress of this large transportation infrastructure project is covered by Wanjohi Kabukuru's blog LAPSSET Tracker: http://lapssettracker.blogspot.de/ (Kabukuru 2012).

[24] See Annex C for two maps depicting THIP's linkage to the transnational transport corridor from Cape Town to Cairo and the LAPSSET corridor that connects land-locked countries in Kenya's North with the Indian Ocean.

[25] Outlined in the new Constitution (GOK 2010: Ch. 11, pp. 108, 113-114) and implemented after the last elections in March 2013, the former county council system was reformed into a devolutionary system of counties headed by governors, which resulted in the new Nairobi County and the new Kiambu County, which includes the former Thika District with Thika County Council and the

While the Nairobi Metropolitan Region is likely to increase its population until 2030 up to estimated 15 million people, much of this massive growth will be found in the peri-urban settlements around the city of Nairobi (Consulting Engineering Services & Runji & Partners 2011: 6-7). Most relevant for the thesis are the towns of Ruiru, Juja, and Thika along the THIP corridor: Ruiru's population is forecasted to increase from approximately 238,000 (2009) to 974,000 (2030); Juja is likely to experience a population growth from roughly 40,000 (2009) to 165,000 (2030); and Thika's population is predicted to increase from approximately 140,000 (2009) to 570,000 (2030) (ibid.: 6-7). This means that the population in these three towns is forecasted to be four times larger only twenty years from now!

Telling from the few sources that deal with a characterization of the peri-urban areas along the THIP corridor, their economic composition features small-scale manufacturing, agriculture and horticulture (particularly tea and coffee farms, dairy farming, crop production, and floriculture), small mining activities, textile and other small processing industries, various (informal) micro-enterprises, and a huge variety of different services and daily life supplies (Municipal Council of Thika: 5; informal meetings with experts in Kenya; field visit notes).

In order to better characterize the study area, field visits were conducted from the end of 2012 until early 2013. During these visits, photos were taken and later organized into categories. In addition, notes on impressions, observations, and conclusions were taken in a research diary. While a discussion on these methodological approaches can be found in chapter 4.2, some insights shall be provided here on the in-situ research experience. In Annex D, a summary on the photographic analysis can be found.[26]

When visiting the peri-urban areas along the THIP corridor, the vast open land becomes apparent as the dominant feature in this corridor. Agricultural uses and livestock breeding can be found in many places. Only a couple of smaller settlements are between the three towns of Ruiru, Juja, and Thika. The overall development of this area is scattered and often no clear development pattern and settlement structure can be identified. So called access or feeder roads to housing plots are earth roads – even in the larger towns, tarmac roads are relatively recent and still very rare.

Even though the area appears not very 'busy' in comparison to the city (or better metropolis) of Nairobi, one can observe a certain dynamic in the peri-urban settlements. Construction is going on everywhere – for both residential and commercial uses. The growth of existing settlements into the open land is apparent. Various billboards advertise the latest, commenced, or soon-to-be-built real estate projects in the area. Also individual land owners are putting up signs to offer their plots for sale or their houses for rent. It is sometimes difficult to figure out or to imagine for the new real estate projects where the displayed 'dream of a lifestyle living' is going to be placed in the vast open land of the peri-urban areas.[27] It is especially the apparent lack of infrastructure and basic services that let the area

---

Municipalities of Thika and Ruiru (which form the functional understanding of the Northern Nairobi Metropolitan Region in this thesis), as well as Kiambu County Council, the Municipalities of Kiambu and Limuru, and the Town Councils of Kikuyu and Karuri – all surrounding Nairobi to the North.

[26] A more extensive collection of these photos in high resolution can be found on the author's website: http://renardteipelke.wordpress.com/research/thika-highway-improvement-project/photos/ (Teipelke 2013b).

[27] See Annex E for a map of selected real estate projects in the Northern Nairobi Metropolitan Region.

appear underdeveloped and not yet opened up for a somewhat guided urbanization process.

The mentioned dynamic can also be experienced in the market areas along the road, particularly in those zones that are described as (possible future) growth nodes – intersections where existing settlements are placed adjacent to the Thika Highway and another major connecting road into the hinterland. Next to already existing shops, small traders with their kiosks and mobile businesses have set up camp. The motorized and non-motorized traffic flowing through these central points creates lively places of economic and social exchanges.

Nevertheless, the 'superhighway' has obviously cut some settlements in half and destroyed a natural beauty of green open space and rivers through infill. Following standards of an international trunk road, the gradient of the highway had to be limited. Thus, the road now runs through the landscape elevated or on a lower level in some parts. The speed on the Thika Highway is quite high with regular traffic accidents happening (Mathiu 2012; Oduor 2013). The quick-stops of matatus at the central meeting points cause some confusion – this is aggravated by a misuse of the service lanes by drivers going in any direction and cutting from these lanes over into the highway (ibid.).

These infrastructure design and traffic management features are related to non-motorized transport on Thika Highway. While the pedestrian walkway and the bicycle lanes along the road are of decent quality, the crossing of the eight to twelve lanes poses a significant challenge. The later installed footbridges are at a distance of up to one or two kilometers. As a result and partly due to habit, pedestrians are jumping onto the highway in order to cross where they need to go.[28]

A detailed description of the various settlements along the THIP corridor goes beyond the scope of this thesis. Therefore, the following unedited extract from the research diary is given as an example:

> "When visiting Juja, I could experience a town that was not a village any more, but also not completely urbanized. Maybe it is a place that most correctly is explained as peri-urban with a tendency to ongoing (but slow) urbanization. It featured many amenities one can find in cities, but then again it has no urban structure. Buildings are standing rather loose next to each other, without a clear development pattern that could be identified. There are various vacant lots, goats and other animals grassing in between, and not much activity going on."

With this introduction to the peri-urban areas along the THIP corridor, the overview of the THIP and its context, as well as Kenya Vision 2030 and Nairobi Metro 2030 is complemented. Following from this chapter on the case study's background as well as the research perspectives of the theoretical chapter, the research design underlying this thesis can be introduced.

---

[28] See Annex D, sub-heading "Crossing", for photos showing pedestrians crossing Thika Highway.

4. Research Design

> *"Ethnographies may lack the apparently 'concrete' results of other methods (with hypotheses proven or not), but an honest and serious engagement with the world is not a failure because it admits that things are messier than that and tries to think through the various complexities and entanglements involved rather than to deny them." (Crang & Cook 2007: 208)*
>
> *"You can explain to them [locals] your research topic, but honestly: Who expects anyone to understand why you want to study a highway!?" (Teipelke 2013a)*

Photo 4.1: Maasai walking along Thika Highway with his goods (Source: Author).

4.1 Hypotheses Based on the Conceptual Framework

In the most general terms, this thesis seeks to study the outcomes of the THIP in the Northern Nairobi Metropolitan Region. As the previous chapters have shown, there are many hidden factors involved in studying a large transportation infrastructure project. First of all, the analysis of this project requires one to open up the 'black box' THIP and to look closer into its idea, its design, its implementation, and then its outcomes. These four elements are all interconnected, build up on each other, and will thus be analyzed chronologically. In line with the argumentation of this thesis, the THIP is understood as a large transportation infrastructure project that was not created in an empty space. Therefore, the project's elements will be discussed in light of the THIP's strategic connection to Kenya Vision 2030. In addition, the analysis of the THIP should be presented within Kenya's broader development framework; i.e. this project is seen as emerging from a distinct political-historical background of how politics and planning have been done in the country. As illustrated in figure 4.1, the five research perspectives introduced in the theoretical chapter of this thesis provide the critical research themes towards which the idea, design, implementation, and outcomes of the THIP will be analyzed as part of Kenya Vision 2030, embedded in the politics and planning context of Kenya.

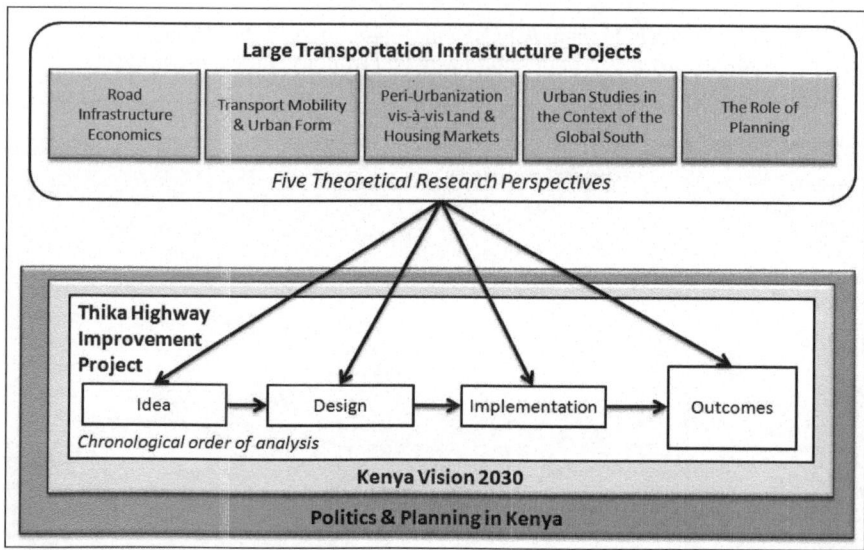

Figure 4.1: Conceptual framework for the chronological analysis of the idea, design, implementation, and outcomes of the THIP as part of Kenya Vision 2030, embedded in the politics and planning context of Kenya (Source: Author).

With regard to road infrastructure economics, the section on the empirical analysis of the THIP and its outcomes will address the debated trickle-down effect which is often ascribed to large transportation infrastructure projects. The interrelation between macro-scale objectives and micro-scale needs will be discussed. Likewise, the correlation between infrastructure construction, economic growth, and social development will be analyzed. The integration of the THIP into the road network and broader planning of the Northern Nairobi Metropolitan Region will be examined. The idea and design of the project will be viewed in relation to different stakeholders' interests.

Concerning the theoretical input from transport mobility/urban form literature, the THIP's outcomes will be evaluated against its design – focusing on the context-sensitivity of this project towards existing settlements along the transport corridor. The use of the road as well as the use of its adjacent land needs to be assessed in consideration of specific interests in order to check for biases in designing and implementing this project. It will also be argued that certain decisions early on in the creation of the THIP have triggered path dependencies resulting in somewhat predictable outcomes. Furthermore, the concepts of mobility and accessibility are key in understanding the differing outcomes of the THIP for various stakeholder groups.

Integrating critical aspects from the perspective of peri-urbanization into the analysis, outcomes of the THIP on the land adjacent to the Thika Highway will be analyzed with respect to peri-urban characteristics of land changes from rural to urban uses and challenges of regional governance related to infrastructure provision. The interrelation between the formal and the informal will also be discussed with regard to effects on residential and commercial uses along the corridor area during the construction of the 'superhighway' as well as after its opening.

With the THIP being interwoven into Kenya Vision 2030, this large transportation infrastructure project also features many aspects that can be critically assessed in light of the urban studies literature that has been applied in the context of the Global South. This means that actors and interests behind the THIP need to be analyzed. This thesis examines the extent to which the city's striving for global competitiveness relates to matters of infrastructure provision and political priorities in projects such as the THIP. The process of developing and implementing the THIP needs to be uncovered in order to critically discuss aspects of participation, representation, and responsibilities. This is related to policy issues of the THIP that span local administrative boundaries and are thus interlinked with questions of regional governance. Outcomes of the THIP will also be studied considering the effects on the spatial segregation on socio-economic grounds, which is deeply ingrained into the Nairobi Metropolitan Region. This aspect highlights the relevance of assessing 'beneficiaries' and 'losers' of the THIP both in economic and social terms.

With regard to the fifth theoretical perspective on the role of planning, the THIP and its idea, design, implementation, and outcomes will be viewed as exemplifying particular characteristics of planning in Kenya. Referring to the research literature, it will be argued that seemingly technical, apolitical planning elements in the THIP need to be (re-) politicized. It will be examined how far planning impacts on the execution of the THIP. The style of government with respect to non-/ integrated planning and the lack of coordination and cooperation along horizontal and vertical lines of government are relevant here. In addition, the role of public and private actors needs to be discussed vis-à-vis the issue of infrastructure provision in land and housing development.

While the above-mentioned critical themes of the five theoretical perspectives have already been presented in figure 2.1, it can be summarized that the empirical analysis of this thesis will show how the outcomes of the THIP are rooted in the Kenyan state regarding the manner in which politics and planning are practiced. It will be argued that the outcomes of the THIP have been influenced by how the project was thought of, designed, and implemented as part of Kenya Vision 2030. For this analysis, outcomes beyond the anticipated project impacts as outlined in the appraisal report (ADF 2007: vi-vii) need to be studied. Furthermore, this analysis demonstrates the way in which the applied theoretical perspectives help to uncover elements of the THIP and aspects of its outcomes that would not be discussed if the THIP were to be evaluated as an isolated project in an empty space. It is the location of this large transportation infrastructure project in a distinct peri-urban area that results in characteristic outcomes. Therefore, the empirical analysis will show that the THIP's outcomes are ambivalent – partly living up to the expectations of the project's proponents, but not resulting in shared benefits for all concerned stakeholders due to flaws in managing the development in this peri-urban transport corridor area, thereby fitting into the Kenyan context of politics and planning.

Before the methodology for this empirical analysis is presented in the next chapter, two limitations of this study must be mentioned. First, the research focuses on the construction phase of the THIP from the end of 2008 until the end of 2012 and the following opening phase until mid-2013. Therefore, the empirical analysis of the THIP's outcomes refers to those changes that are already apparent in the study area and some changes that are likely to develop in the future. For that reason, the findings are not meant to be definite and irrevocable. Future developments will show the extent to which outcomes of the THIP are short or long-term and whether they are influenced by external factors. Some of these factors will be discussed in the empirical analysis.

The second limitation to this study is in relation to the research area. Since the specific importance of peri-urbanization in metropolitan regions of the Global South – as exemplified in the case of Nairobi – has been identified in the theoretical chapter of this thesis, not all three sections of the Thika Highway that were part of the THIP will be analyzed in great detail. The focus will lie on the third (largest and longest) section of the THIP (see figure 1.1). This area starts at Kenyatta University, passes the towns of Ruiru and Juja, and goes up to Thika and beyond. The peri-urban character of this area makes it of particular interest to this study. The other two sections will necessarily play a role in the empirical analysis, since most outcomes of the THIP cannot be solely assigned to certain parts of that transport corridor. Nevertheless, there will be a lesser focus on land and spatial aspects in the two sections that lie in the city of Nairobi due to the developed and urbanized character of these areas.

4.2   A Strategic Mix of Methodological Approaches

Based on the above outlined hypotheses, a set of methodological approaches was applied to adequately analyze the THIP. Figure 4.2 presents the various sources of input into the research work. They are organized in three categories, covering desk work, field work, and the mixed desk-/field-work task of interactive work. The items in the illustration had different levels of importance in the final analysis.

With regard to the desk work, an in-depth research of both primary and secondary literature was the key theoretical foundation of the thesis. This task was supported by a search for newspaper articles. The corresponding articles were used as a source of information to understand the 'story' of the THIP and how it was referred to in the public debate.[29] The video of a discussion forum on the THIP (held 20 November 2012 at the University of Nairobi) was useful in prioritizing sub-themes in this thesis' research. The 'makeshift' comparison of Google Earth satellite images of the study area provided some information on spatial changes along the Thika Highway from before until the end of the THIP implementation. These findings were compared to the few research articles that deal with satellite image studies of the Nairobi

---

[29] Relevant exemplary newspaper articles will be referred to in the analysis of this thesis. Information on the research for newspaper articles and comments on the search results are presented in Annex F. A complete list of the found newspaper articles can be found on the author's website (Teipelke 2013b). Going beyond this analysis, it can be remarked that a discourse analysis of these newspaper articles could reveal more information about what role language plays in framing the THIP in a performative way.

Metropolitan Region (Mundia & Aniya 2006; Mundia & Murayama 2010). The role of the research diary went beyond the practical function of covering the research progress and following up on open tasks, in that it provided the opportunity to put the empirical material into context and helped to regularly call for a reflection of the author's multiple roles in the field. Annex G provides a discussion on the issue of positionality of the researcher as well as four extracts from the research diary exemplifying the field work experience during this research project.

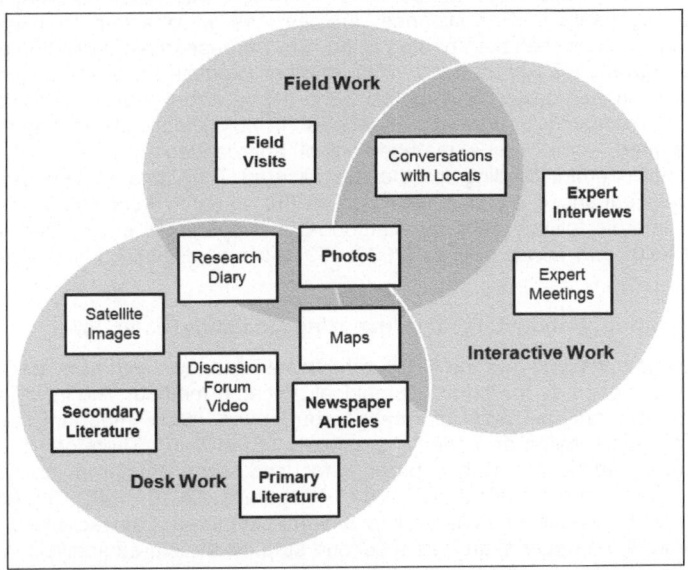

Figure 4.2: Methodological input into the research project categorized in field, desk, and interactive work, with the most important items in bold (Source: Author).

The second major basis for this thesis were expert interviews (further discussed in chapter 4.3). They were complemented by informal expert meetings, in which ideas were exchanged, research observations were discussed, and advice was given on how to best proceed with the field and interview work. These two elements of the interactive part of the research work were accompanied by conversations with locals in the study area in order to learn more about their perspectives on the THIP and their lives in the peri-urban areas ("receptive interview"; Lamnek 2010: 340-348).

The third key element in this research work were several visits to the study area by matatu, car, and primarily foot. These field visits were not a participatory observation, but rather an active and direct engagement of the researcher with the THIP and some of its outcomes through the researcher's role as a transport user (both driving and walking) and market/shop customer.[30] During these field visits, photos were

---

[30] In Annex H a reflection on this field work experience can be found in an article that was originally written for a professional blog (Teipelke 2012a). This article refers to authors like Sidaway (2009), Wylie (2005), and Paasche and Sidaway (2010), who have written about the methodological technique of consciously walking through a study area in order to discover complex geographies both with regard to the details of a 'research object' and the relation between the researcher, the researched, and the space in which this interrelation develops.

taken, organized into categories, and then analyzed in comparison with in-situ observations and desk work findings (Harper 2012: 402-416).[31] In addition to these methodological inputs, maps were used in both the desk and field work, as well as in conversations with experts and locals to compare what the maps were depicting and what could actually be found in the field – or as Bayfield frames it: "Walking the territory redraws the map" (2009).[32]

Various methodological approaches were applied in this research work and they were also overlapping: For instance, interviewees were asked to comment on recurring motives on the taken photos or on how the researcher himself experienced the Thika Highway (cf. Harper 2012: 414). Another example are the field visits, during which locals in peri-urban settlements were asked about THIP stories that were found in the research for newspaper articles. In conclusion, the research material was developed from various methodological approaches in order to facilitate a comprehensive understanding for the multifaceted THIP and related topics. This strategy also enhanced a critical reflection on the research work and aspired to live up to the interdisciplinary calls for methodological openness exemplified in papers such as Phelps and Tewdwr-Jones (2008: 579) that was discussed in chapter 1.3.

4.3 Preparing, Conducting, and Analyzing Qualitative Interviews

In order to gain insights into the THIP, qualitative interviews with relevant experts in Kenya were chosen as the primary empirical research method. The interviews were thought to not only provide information, but, even more important, to provide reasoning/argumentation on underlying aspects of the THIP. Therefore, the method of qualitative interviewing with a problem-focused, semi-structured, questionnaire-based style was applied (Helfferich 2011: 45; Lamnek 2010: 350). Conducting the interviews with a questionnaire and in a semi-structured way facilitated both the organization of the material and, as a second step, its structured analysis. Instead of using a completely structured, strictly questionnaire-following interview technique some narrative elements were included in the course of the interviews, in which the interviewees could talk about THIP elements or broader aspects that were not a major part of the research scope. In addition, the chosen technique also helped to adjust formulations and re-order questions during the interviews. This made possible reactions upon individual interview situations and diversions from the typical structure (Helfferch 2011: 42).

The questionnaire was developed from the analysis of the research literature and initial meetings with experts in Kenya. From a catalogue of possible themes, questions were formulated, organized in categories, and ranked after priority/relevance to answering the research questions.[33] After the first two of the 17 interviews, small adjustments were made to the questionnaire and some questions were reformulated, although the existing structure and content remained the same.

As the questionnaire in Annex I shows, the first part of the questions referred to a description of the THIP and its positioning towards Kenya Vision 2030. Thereafter, actors and interests of the THIP were discussed. The third part concerned the

---

[31] See Annex D for more information on this photo exercise and a best-of selection of photos.

[32] It is recommended to take a look at the previously shown maps as well as Annex E and then to compare imagined places with the photos in Annex D.

[33] These tasks followed the common SPSS principle – standing for the German words "sammeln, prüfen, sortieren, subsumieren" (collect, check, order, prioritize) (Helfferich 2011: 182-185).

implementation process, while the fourth part turned to the outcomes of the THIP. The fifth part was titled "Global vs. Local", although the questions in that part covered mainly the issue of the project's direction with regard to objectives and targeted scales. Questions in this part were often asked as part of other discussions during the course of the interviews. In the sixth part of the questionnaire the comparability and the idea of the THIP as well as decision makers' understanding of local needs was thematized. The last part referred to conclusions on the THIP by the interviewees. The empirical analysis will show that this organization of the questions was partly helpful in structuring the material. Nevertheless, it was the particular analysis afterwards that revealed most of the insights and the relations between various aspects, causing a reorganization of the empirical material.

In relation to the selection of interviewees, a list of possible contacts was compiled and further extended through 'snowball sampling'[34]. The aim was not to perfectly represent every possible stakeholder group of the THIP, but rather to gain insights from different perspectives by interviewing people that brought with them a certain expertise and experience on the THIP and related topics (such as politics and planning in Kenya) (cf. Lamnek 2010: 350-353). People interviewed or talked to qualified for this empirical work, because they either can be considered as having been directly impacted by the THIP, having been part of the THIP and the institutions behind it, or having worked in the relevant professions that are concerned with the THIP and the specific study area. This means that interviewees were chosen from national and local governments, including bureaucrats and consultants (normally working temporarily for public institutions in specific programs); representatives of bodies representing a certain stakeholder or professional group; experts from international agencies who are working in infrastructure and urban development; actors of the private sector; journalists who have been reporting on the THIP for several years; university researchers (who often also worked as government consultants in the field of infrastructure and urban development, as well as land and housing issues); and locals (both from a residential and commercial background).

The list below presents the selected interviewees with information on their professional background qualifying them as interviewees for this research. When referring to a specific interviewee's comment later on in the empirical analysis, the anonymized description given in the list will be used (for instance: interview journalist A). The list shows 18 interviewees, of which two participated in a double-interview, thus the overall number of interviews was 17. Two informal meetings were included in the empirical research and conversations with twelve locals were conducted during the field visits.

Interviews:
Local government bureaucrat: responsible person for infrastructure works in one of the counties in the THIP area
Local government consultant: urban development, safety, and shelter expert with experience in monitoring and evaluation

---

[34] The term snowball sampling describes a technique that can be applied in interview work in that contact to further interviewees is facilitated or enabled by previous interviewees through their personal relations. It helps to reach persons who would otherwise be difficult to be contacted or whose relevance or expertise could not be known to the researcher beforehand.

National government bureaucrat A: urban development expert in the Ministry of Local Governments

National government bureaucrat B: urban planning expert in the Ministry of Nairobi Metropolitan Development

National government consultant (and professional body representative): urban and regional planning expert and researcher with experience in local government capacity building

Civil society organization representative: participatory planning expert of a key stakeholder group in the THIP

Professional body representative (and government consultant): land surveyor and spatial planning, land management, and capacity training expert

African Development Bank expert: transportation infrastructure specialist

UN-Habitat expert A: land use planning expert with long-term experience in Kenyan politics

Investor: larger market actor with big investment along the THIP corridor

Real estate agent A: expert in upper-market land and housing with larger real estate project along Thika Highway

Journalist A: editorial journalist at the Daily Nation

Journalist B: freelance journalist who has closely followed the THIP for several years

Journalist C: journalist and land economist at the Standard

University researcher A: public policy, land use, and displacement expert who has conducted research on politics and planning in the Nairobi Metropolitan Region and the THIP for several years

University researcher B (and government consultant): urban and regional development expert with corresponding research and government-related consulting work on Kenya and Nairobi in particular

University researcher C (and government consultant): urban and transportation planning expert with corresponding research and government-related consulting work on Kenya and Vision 2030-related infrastructure projects in particular

University researcher D (and government consultant): real estate and land expert with corresponding research and government-related consulting work on Kenya and Nairobi in particular

Informal Meetings:

UN-Habitat expert B: land and governance expert for Kenya

University researcher E: East African Community and Kenyan party politics expert

Field Conversations:

Three local students: young Kenyans who are both studying and living next to the Thika Highway in a peri-urban settlement

Three local pineapple sellers: operators of mobile stands along access roads to Thika Highway

Three local grocery shop employees: workers in a small grocery shop in a peri-urban settlement along Thika Highway

Local pharmacist: business owner in a peri-urban settlement along Thika Highway

Local manufacturer: regular customer in the wholesale business along Thika Highway

Real estate agent B: representative of a cooperative which has a large-scale upper-market real estate project in the Northern Nairobi Metropolitan Region

In order to conduct the interviews in a proper methodological way and to respect ethical research standards (Helfferich 2011: 190-192), the background of the research was always explained to the interviewees – often during the first telephone conversation. The interviewees were informed about the use of the interview, which was recorded and later anonymized for the presentation. The decision on the anonymization was based on two considerations: Many interviewees seemed more willing to openly talk about political hot-button issues with the knowledge that the interviews would be anonymized. In addition, the provision of interviewees' names does not provide any significant further information for the empirical analysis of the THIP (cf. Lamnek 2010: 352). All interviews were anonymized, even though a few interviewees expressed that they would have no problem in being quoted by name. The interviewees were informed that they receive an electronic copy of this thesis once the research is finalized.

As part of the research diary, notes were taken after each interview to reflect on impressions and observations. This 'postscript' was used as a safeguard against the researcher's own preconceptions and their possible impact on the research (ibid.: 335; Helfferich 2011: 52). At a later point, the notes also helped to put the empirical material in the right context. With the software "f4", complete transcripts of the interviews were done based on audio records. From memory, transcripts were done from informal expert meetings and conversations in the field immediately after corresponding events. The transcripts included phonetic, facial, and gesture expressions – even though they might not have been of use in the final analysis (cf. Mayring 2010: 12, 2002: 91-94).

Following the steps of analysis outlined by Mayring (2010: 93; see Annex J), a category system for analyzing the interview material was developed. For the actual analysis the software "MAXQDA" was used. The first category system (or code tree) was developed out of the interview questionnaire and the in-depth literature research, in which relevant aspects for the analysis were identified. Aiming for a qualitative structured content analysis of the interview material, some of the initial codes were merged or reformulated. These adjustments did not change the direction of the analysis, but were helpful for the organization of the material. During the analysis, some material called for additional codes to be added to the category system. By going through the material twice, it was ensured that the final category system was equally applied to every interview's content. The smallest piece of material that could be coded was defined as a single word, and the largest piece of material that could be coded was defined to be a whole paragraph.

The complete code tree (category system) can be found in Annex K. The six major categories for analysis are partly similar to the interview questionnaire. In the first category group, material was coded that contained descriptions of the THIP and its idea. In the second category group, the project design with regard to actors and processes was analyzed. The third category concerned the implementation of the project referring to direct results of the THIP *during* construction and the underlying reasons for them. The fourth category group covered outcomes of the THIP in terms of economic, political, provisional, social, and spatial aspects. One further sub-category referred to the time dimension of these outcomes, and another sub-category to winners and losers of the THIP. In the fifth category group, material was coded that provided information on underlying reasons, broader issues, and the overall framework in Kenya. This material was organized along different perspectives – depending on the reference to, for instance, the historical context, institutions, politics, role of various actors, or spatial characteristics. The sixth category group included categories for material that provided insights beyond the project implementation in that comparability, targeted scales for the project impacts, and concluding 'take home aspects' were coded.

Altogether, 1,809 coded elements were analyzed. The multiple coding of some parts of the material was not thought to pose problems with respect to methodological rigor. More material had to be analyzed; however, fundamentally important interview passages were not overlooked. For this second step in the analysis, most of the material was summarized in paraphrase and few distinctive parts used as direct quotes (Mayring 2010: 70, 98; Lamnek 2010: 473). As the chapter on the empirical analysis will show, the material was reorganized/regrouped after this systematic content analysis for the final presentation. The reason for this was that material under some sub-categories proved most fruitful for the analysis, when it was put in dialog with material of other sub-categories (Mayring 2010: 94). Furthermore, due to the limited space of this thesis, only those parts of the material were eventually used that referred to the above-outlined research focus and the corresponding hypotheses.

4.4 The Challenge of Data

The last section of the methodological chapter is dedicated to one of the single most important issues for conducting research and working in practice in Sub-Saharan Africa (and many other parts of the world): the existence, the access, and the quality of data. Various authors have underscored that data in Sub-Saharan Africa is often incomplete, non-existent, out-of-date, forged, not sufficiently detailed, unreliable, modified, and/or based on unclear assumptions (Rakodi 1997b; Ayogu 2007: 118; Simon 1997: 55). This makes research on a topic in only one particular place difficult and comparisons between places and cases a daunting task (Simon 1992: 13-14).

This data challenge is directly related to practical politics in many cases. There are plenty of cases, in which researchers cannot access the necessary information for their research from government or other sources (interview university researcher A, C; Becker 2011: 11). On the other hand, the exact same institutions might experience a similar problem of lacking relevant data to properly manage their work fields and implement policies as planned (Dowall 2003: 15, 18). In practice, reference to market solutions are, for instance, rendered irrelevant if market participants are in a situation where they cannot obtain the information that is necessary to make informed decisions (cf. Grant 2009: 142). This whole issue raises questions of transparency and access to information for both research and practice that will be taken up in the empirical analysis.

While these challenges partly had to be dealt with in this thesis' research, the strategy of 'triangulation' was thought to balance out this problem (Flick 2012: 310-315; Lamnek 2010: 245-265). Of the various elements in the triangulation strategy, the use of different sources and types of data, the reference to different theoretical research fields, and the use of several methodological approaches was applied to develop a comprehensive understanding of the THIP despite some difficulties with regard to the access and quality of data.[35]

Since this research project aims at addressing this challenge of data, this document contains extensive information in the appendix. Furthermore, a website provides background information and additional material (Teipelke 2013b).[36] This does not only address the issue of research ethics regarding publicly sharing information and knowledge (Sidaway 1992: 406), but also gives other readers the opportunity to interpret the sources of this thesis differently. This is an aspect pointed out by Pieterse (2010b: 23) who notes that it is not only the availability of data but also its transparent interpretation that decides over the quality of the conducted research.

---

[35] See Annex L for a discussion of the quality of qualitative research.

[36] As one component of the author's website, extensive information and (additional) material on this research project is provided for a broader audience online: http://renardteipelke.wordpress.com/research/thika-highway-improvement-project/ (Teipelke 2013b).

5. Empirical Analysis

> *"The perceptions and reactions of citizens who are seeing their lives transformed by the highway have not been part of most discussions about the highway, and little independent research has taken place on the impacts of the highway especially for non-motorists." (KARA & CSUD 2012: 2)*

Photo 5.1: Food stands and boda boda at matatu stop in Witeithie (Source: Author).

The presentation of the results of the empirical analysis will start with a discussion of the idea of the THIP, followed by an analysis of this project's design and implementation. Then, the outcomes of the THIP will be presented with regard to transport and economy, as well as land and housing. In the final part of this chapter, conclusions will be drawn on how the empirical analysis links up with the previous chapters of this thesis. As explained before, critical themes of the five research perspectives will now be applied to the chronological analysis of the THIP's idea, design, implementation, and outcomes.

5.1 The Idea of the THIP

In order to analyze the THIP and its outcomes, the roots of the project have to be studied. For this it can be revealing to understand how stakeholders actually describe this large transportation infrastructure project. While on paper this project is basically an expansion of a four-lane road into an eight-lane highway, people interviewed for this thesis see very different things in the THIP. Their description of the project ranges from the local to the metropolitan, to the national, and even to the trans-national scale, seeing the THIP either in isolation or as part of a larger engagement by the Kenyan Government.

One group of interviewees described the THIP as a highway upgrade in reaction to two challenges: the heavy congestion in that transport corridor and the economic potential that was inhibited by the severely limited mobility as well as access in this area. The mobility aspect is seen in relation to the difficulties in movement on the road – for both goods (export-import, metropolitan) and people (farmers/market traders, commuters). The access aspect refers to the peri-urban areas in the Northern Nairobi Metropolitan Region that were seen to be calling for an opening up by appropriate road infrastructure in order to enable economic growth. Therefore, this interviewee group is closest to the 'road infrastructure-economic growth' causality view that was presented in chapter 2.1. Its simple causality model can be illustrated by several interview quotes[37]:

> "But remember that part of Vision 2030, job creation is a very important thing (…). And how do you create jobs, without having an efficient infrastructure? (…) This simply means the Government really wants to (…) do like improving infrastructure and then people can do business." (interview journalist C)

> "(…) If you afford people some level of mobility, you afford them some levels of economic improvement at the absolute level. You don't if you're not going to move, you can't improve your livelihoods you know." (interview African Development Bank expert)

> "The implementation of Vision 2030, ah, construction of infrastructure, roads being part of it, is very key. Because it's the engine that will move all other sectors, including social sectors, towards Vision 2030." (interview university researcher C)

The critique that can be found in the transport mobility/urban form literature against such a simplified view is also apparent in the answers of another group of interviewees: These experts described the THIP as an isolated road infrastructure project that lacked context-sensitivity from the beginning. Neither socio-economic, nor environmental, nor planning considerations had been taken into account sufficiently. Even though the African Development Fund financed a study on public transit along this transport corridor, the intended rapid transit system has not been realized. Moreover, this group of interviewees often highlighted early on in the interviews that non-motorized transport uses as well as public transit have played no or only a minor role in comparison to private car use. Besides, a few interviewees raised the topic of the railway that could have also been upgraded/extended as part of this infrastructure investment in the Northern Nairobi Metropolitan Region

---

[37] Also illustrative is the rather simple reasoning in the THIP appraisal report: "The project is expected to contribute to enhance transport services and urban mobility in the Nairobi Metropolitan Area by reducing general transport costs, improving accessibility to public transportation, employment opportunities, housing, and recreation activities. In addition the project is expected to promote private sector participation in the management and operation of road infrastructure in Kenya. *The project will therefore have significant impacts on poverty reduction in Nairobi Metropolitan Area*" (ADF 2007: 19; emphasis added).

(interview local government consultant, professional body representative, university researcher B).

A different description of the THIP is provided by those interviewees that put the idea for this project in a political-historical context, as the interviewed national government officials did. On the one hand, this project was born out of the Nairobi Metropolitan Growth Strategy of 1973 that identified growth areas in which certain infrastructure projects would become necessary in order to accommodate population increases and spatial expansion. One government official described how the Kenyan Government tried for many years to acquire international financial support for this road project, but did not succeed, since the country was isolated under the late tenure of the second President, Daniel arap Moi (1978-2002), in the 1990s until 2002. With the third President Mwai Kibaki (2002-2013) coming into office, infrastructure upgrades were put high up on the agenda and international donors at least partly adjusted their policies towards Kenya. Nevertheless, the THIP started only after the African Development Bank and the Chinese Government could be won as financial supporters, since other institutions and countries showed no sufficient interest in supporting this project.

In addition to this political-historical context that has an international component to it, some interviewees also discussed an internal political feature of this project: the fact that the location and direction of this highway project was towards the region from where President Kibaki was from. As one interviewee illustrated:

> "Kenyans always think presidential race is that matter of life and death. They assume we choose a President; in fact now, people don't do really think that Thika Highway had been rationalized. They simply think because we elected a President. That's their notion. That's why they also want a President, so they could also have a highway like Thika Highway." (interview university researcher D)

This 'logic' behind the THIP could be rebutted in the empirical research for this thesis, as the project's dimensions were far too great and the overall necessary political support included a broad coalition. And as was just described, the project's roots were already decades-old. Nevertheless, this political connotation – even though it might be an unjustified assumption – plays a role in the THIP, as one interviewee explained:

> "In many nations, infrastructure has got a lot of political connotation. Actually, recently, towards the end of last year, there were lot of complaints from other MPs [members of parliament] from other regions. What they were asking: How can we use billions to construct a road of 50 kilometers, if in other places there's no access?! Why should we not have developed a road of moderate standards and we distribute that money to be able also to access, ah, to construct roads and/or improve other roads in other places where access is not?! Why?! And they were associating it with the fact that the President is from this [region]. So what I'm telling you: There's no place where you talk about infrastructure without talking about political connotations. Actually, very many people are still feeling aggrieved, even up to now, how can we use over 30 billion to do one road of 50 kilometers?! (...) So I can tell you some of these decisions, people always see political connotations. Maybe there

aren't, but people will always see, ah, a citizen will always see political connotations. And not only roads, but any form of infrastructure." (interview university researcher C)

This political connotation that is part of the broader perception of the THIP – both in everyday conversations, in the media (see for instance Odipo 2013, Pala 2012, Wamwere 2013, Geita 2012), as well as in the interviews conducted for this thesis – has to be kept in mind, when the project's design, implementation, and outcomes are discussed.

Referring to the THIP's role with regard to Kenya Vision 2030, the THIP was not seen as a flagship project, although interviewees confirmed its clear connection to the vision. This means that the THIP was not seen as a special project, but as one part in the implementation of Kenya Vision 2030. In the interviews, many experts made a very quick connection between this road project and the overall government investment in infrastructure through the guiding vision. Some interviewees emphasized the transport element in Kenya Vision 2030 and others saw infrastructure investments, such as the THIP, in relation to economic growth, while another group of interviewees interpreted the role of the THIP in light of its symbolic value:

"It can be seen as a major element (...) – one aspect of it is the symbolism involved in Vision 2030. The country wants to be seen to be doing projects, or to do projects with big impact. And a huge aspect, (...) kind of like, ah, big iconic projects. And therefore, this was one indication that the country is capable of doing that kind of project." (interview university researcher B)

This reference to the symbolic element of actually implementing 'big projects' changes the perspective on the THIP from solely being a technical infrastructure project to one that also has immaterial features, which are relevant to the objectives and implementation of Kenya Vision 2030. Here, the road infrastructure and planning literatures come into dialog with the urban studies literature as this project embraces different functions of, for instance, improving traffic, following an implementation plan of Kenya Vision 2030, and showcasing the country's development will as it is outlined in Kenya Vision 2030. At this point, the opening up of the 'black box' THIP can provide more detailed insights by studying the project's design. Before this is done, the above discussed elements of the THIP's idea are illustrated in figure 5.1 vis-à-vis the relevant theoretical perspectives and their critical themes that were summarized in chapter 2.6.

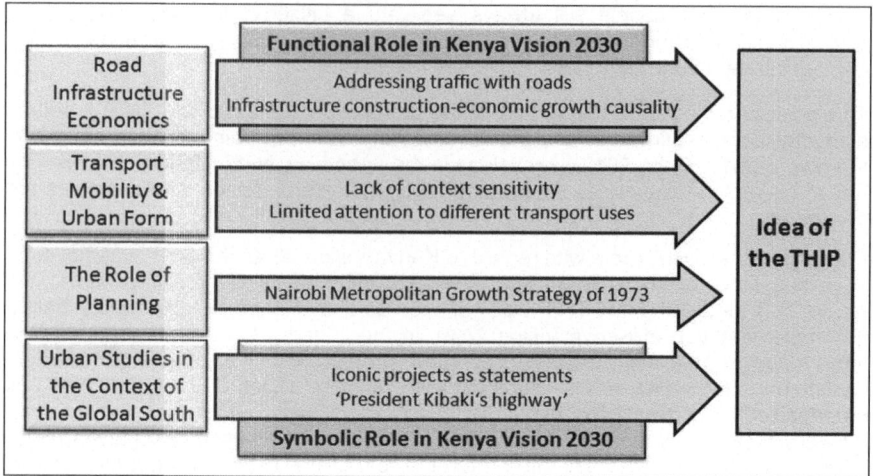

Figure 5.1: The idea of the THIP as part of Kenya Vision 2030 vis-à-vis relevant theoretical perspectives and their critical themes (Source: Author).

5.2   The Design of the THIP

In order to trace back the design of the THIP, two elements need to be analyzed: the actors in the THIP design phase and the actual process of designing it; i.e. answering the 'who' and 'how' question of the project's design.

5.2.1   Actors in the THIP Design Phase

Identifying who actually worked on the THIP is a difficult task. This has to do with the fact that available documents as well as the information given in the interviews resulted in a long list of several actors that had somehow been involved in the THIP. On the other hand, this is not too surprising, since a project of that scope and scale automatically affects several different stakeholders that are included in the design of this project to a varying extent. Without already turning to the process, it can be remarked that most of the stakeholders were not sufficiently involved in designing and coordinating the THIP and related activities (KARA & CSUD 2012: 2, 11-13; Earth Institute 2012; Klopp 2012a).

As was already described, the THIP had its roots in the 1970s and was revived and eventually activated under Kibaki's Presidency – with the President himself and Prime Minister Raila Odingo really putting a lot of political effort behind the project. The Ministry of Roads, which was headed by the Prime Minister, had the responsibility for the THIP, while one of its 'outsourced' authorities, the Kenya Urban Roads Authority (KURA) managed the actual implementation. Interviewees' answers to the question about actors referred to seven ministries that were described to had somehow been involved in the THIP (finance, transportation, water, Nairobi Metropolitan Development, local authorities, energy, lands); however, it became apparent that interviewees were often assuming on how the involvement should have been – not how the involvement actually was. The THIP was designed in a vertically and horizontally highly fragmented government system. Other national ministries did not play a very active part in this project and smaller-scale authorities, such as the affected counties in Nairobi, Ruiru, and Thika, were also rather 'informed' about the project instead of having had a significant and lasting input into the design of the THIP.

This view on actors needs to be extended, particularly to the other project financiers. The African Development Bank had its input into the design of the THIP at least to some extent, since the original plans were checked against the bank's safeguards for infrastructure projects. This meant that recommendations were made with regard to pro-poor, gender-aware, context-sensitive designs. As the discussion of the implementation of the THIP will show, this input did improve the project's design, but eventually it was the Kenyan Government that decided how and to what extent these recommendations would actually be implemented (interview African Development Bank expert). The particularly recommended, but not included bus rapid transit system – for which even a study was conducted – tells something about this aspect.

While most interviewees did not further elaborate on the role of the Export-Import Bank of China and the Chinese companies that constructed the highway, one of the interviewees with the longest experience in Kenyan politics described their role as follows:

> "(…) So, very often, roads networks are not implemented in terms of objective priorities, but who wants which contract and who brings the money to do it." (interview UN-Habitat expert A).

And concerning issues that were raised by the African Development Bank, this interviewee saw the sensitivity of external donors for on-the-ground aspects in a critical light: "They don't care. For them, it's business" (ibid.). A discussion of foreign countries' engagement in Sub-Saharan Africa goes beyond this thesis[38], but the question arises in how far this 'business' aspect hints at broader issues of differing perspectives and interests of participating actors in the THIP – with the Kenyan Government (with ministries and lower-scale authorities), the African Development Bank, and China (including the contracted companies for the THIP construction) looking differently on the THIP. As a result, some actors likely saw the THIP from an international, or at least trans-national economic-trade perspective, others understood it as facilitating transport mobility, and some might have seen this project targeting micro-scale socio-economic objectives. Given the fact that the local private sector, non-governmental organizations in Kenya, as well as the public in general had no say in the design of the THIP (KARA & CSUD 2012: 2; for instance interview civil society organization representative, university researcher C, investor; real estate agent A), it comes down to who has the power and who puts the money on the table to make certain project elements priorities and to put other possible concerns to the side (interview UN-Habitat expert A, university researcher A).

What is really important to emphasize – and this might partly appear to contradict what was just argued – is the fact that there was no outright opposition to the upgrade of the Thika Highway. Many interviews as well as the media coverage underscore that the broad and advertised objective of improving the traffic situation

---

[38] Even though the research for this thesis did not uncover a particularly strong role of the People's Republic of China or its Export-Import Bank in specifically affecting the THIP and its outcomes, one could discuss the THIP in a larger framework of China's enhanced engagement in large infrastructure projects in Kenya and Sub-Saharan Africa in general (see for instance Day 2013; Owuor 2010; Guangyuan 2012; Kwama 2013).

along the Nairobi-Thika transport corridor was never disputed. Therefore, it would be wrong to discredit the THIP as a 'big actors' project disregarding the general public's needs. As one interviewee pointed out for doing projects as the THIP in Kenya:

> "Whatever projects that are undertaken, whether consultations with the public is not done in detail or not, all our projects arise out of necessity...out of an expressed need. An expressed problem that people are seeing. They don't do it because they're copying elsewhere. They're doing it because it is a necessity! I'm saying that I know the problem is the extent of consultation based on the Constitution is not sufficient. But the projects that come in place, they arise as a result of necessity." (interview university researcher C)

One also has to see the THIP and citizens' support for it in the historical context of so called 'white elephants' in Kenya – public projects that politicians pompously promised but never delivered. As the same interviewee explained with respect to road projects in North and North Eastern Kenya, which together with other projects such as the THIP form the road infrastructure package in Kenya Vision 2030:

> "For a place like Kenya where people like in North Eastern and Northern Kenya have not seen good roads – whatever comes as a road, they will never say anything against it. Like if you would go here to interview a normal ordinary person and you talk against this road, they will not understand what you're talking about. Because for them that's the most beautiful thing that has ever happened. (…) Because they can't see the disadvantage. (…) What they see best is the advantage: 'Now, I can move. Within this time, I'm here. Oh, that's beautiful.' They don't look at what is it to do with my land value. What is it to do with my safety. What is it to do with my security. They don't look at that. That they only realize at a later time." (interview university researcher C)

One could add to this observation that people could not really have evaluated for themselves how 'comparatively good' the THIP design actually was, since there were never serious attempts to present alternative options to a public debate. Critique against the THIP arose when the project moved from the planning to the construction stage. And this has primarily to do with the 'how' question of the THIP:

> (…) What I would like to say here is that, ah, the concept of the project has never been, ah, no one has been against it. Nobody has expressed any reservation. But of course, as the project has been implemented, people obviously look at it and say: 'It should have been done a better way. It should have been done in a better way.' The idea that the improvement can take place is not questionable. But the process (…)." (interview civil society organization representative)

Since this judgment is also supported by a key report done by the Kenya Alliance of Resident Associations in cooperation with the Center for Sustainable Urban Development of Columbia University (KARA & CSUD 2012) after in-depth consultations with local stakeholders, several issues of the THIP design have to be discussed by looking at its process more closely.

5.2.2 The Process of Designing the THIP

Assessing the design of the THIP from studying its process includes three relevant aspects that were discussed in the interviews: information access, meaningful consultation, and the integrated project approach.

If the stakeholders' engagement in the design process shall be discussed, the role of information access is key and in the case of the THIP a critical, negative aspect. Several interviewees told their story of how they tried to learn more about the THIP and its design, but could hardly access even the most general information:

> "I have carried out a study on that road. Getting the design has been hectic. Nobody is willing to part with the design. A public document, which is being paid for by the public, nobody would like you to see it! (...) I don't know! Is it that there's something that they're hiding or something?! It became very interesting and to me, (...) I think that many stakeholders felt it in that particular respect." (interview university researcher C)

> "So, the [information] is not available, no public hearings. We could not get the designs. The designs were never released in public. We had to fight. There were some, one person, (...) at the Ministry of Roads, that would allow people to come in and look at the designs (...), but they never put the designs out for people to actually come in and (...) to actually say: 'That is not a good idea.' (...) Overall, the high-level non-transparency that is highly problematic in my view." (interview university researcher A)

Assuming that these experts have gained some experience in dealing with the Kenyan Government, one can imagine how difficult it was for ordinary citizens to meaningfully engage in the design process of the THIP. This is a point where a seemingly technical infrastructure project becomes a political issue and questions of legitimacy and responsibility arise. It has to be emphasized here that the government did deal in a legally correct fashion with people whose property was for instance taken by the highway expansion (see chapter 5.3 on the THIP's implementation). The problem, however, is that the responsible people in government seemed to have been assuming who a 'relevant' stakeholder was and who was not, meaning that the public in general was not considered to be important to engage with this project. This is related to a political culture of doing such projects that will be taken up again in the policy recommendations in chapter 6.1.

With regard to consultation, the THIP showed some improvements in comparison to previous large transportation infrastructure projects, since there were at least limited stakeholder meetings and the input from the African Development Bank resulted in some changes to original project plans (for instance interview local government consultant, professional body representative, civil society organization representative). But the report by KARA and CSUD (2012: 2) outlines several problems with the process and its consultative elements: The outreach to the public was never done in a structured way and lacked sufficient communication concerning information about public events. Furthermore, the project designers did not seek much external advice from other groups, such as academic researchers, residents' associations, business alliances, etc. (also see Ngirachu 2010). When KARA, for instance, was brought in, the implementation of the THIP had already been long under way and then it became a matter of fixing problems. Some interviewees described this style of consultation as 'minimal', where the government is following the law, does not break the rules, but nevertheless lacks the type of meaningful engagement where several project options/designs are actually on the table and open to public debate.

This aspect is related to a severe lack of project integration into other policy areas that resulted in corresponding implementation delays (Choto 2010). The THIP had been linked to other road projects, such as the bypasses for the Nairobi Metropolitan Region, but besides that: "The whole project was the road only, and what happens left and right to the road (...) was left basically to market forces and speculation" (interview UN-Habitat expert A). Some interviewees claimed that there have been plans in place for infrastructural, spatial, or socio-economic development in the project area that were not shared or used due to the silo thinking and sectoral planning style in the Kenyan Government. Other interviewees claimed that information was not there and that other authorities had not done their 'homework' in terms of preparing development plans for their administrative areas.

While a lack of *integration* of policy areas (and plans, if they existed) can already impact on the design and implementation of the THIP, it becomes even more critical, when corresponding government activities are even *dis*integrated. One infamous example is the beautification program of the city of Nairobi that had upgraded areas in the city that were later completely ripped open to accommodate the expanded road infrastructure. Another example in the interviews was the Kenya Municipal Program by the Ministry of Local Governments that was also implemented in Thika and addressed policy fields directly related to the THIP, but had no connection to the highway project:

> "(...) The idea was just to help Thika as a municipal place to upgrade its infrastructure to the standards required. Nothing to do with the highway per se. The highway is incidental to the objectives of the Kenya Municipal Program. Cause the Municipal Program supports planning, supports institutional strengthening, and it also supports infrastructure, roads, drainage, and street lighting, etc. (...) You know the Highway Authority is separate from ourselves. And they do their own things. Perhaps the coordination is not as good as it could be." (interview national government bureaucrat A)

Interviewees with experience both on the higher and lower levels of Kenyan Government complained about the lack of information sharing between different offices. This character of sectoral institutions to "keeping things to themselves" (interview professional body representative) results in a planning system that is not only reactive but also not informed by what different offices are working on in the same policy fields or project areas. In such a fragmented system that lacks a culture of cooperation, long-term planning, investment, and commitment to one project, a specific place, or a particular policy challenge seem hardly possible:

> "(...) we just focus on one aspect and put some money into water, put some money in transport infrastructure, and other, housing as well. So, they will all come up at the same time, but not necessarily being informed by a central thinking, planning, you know." (interview African Development Bank expert)

Interviewees described this situation as emerging from a situation where top policy-makers are taking the responsibility for projects such as the THIP, but local needs are disregarded, because the relevant stakeholders (even local authorities) are not sufficiently consulted (interview national government bureaucrat A, local government bureaucrat). This exemplifies what has been discussed in the theoretical chapter on integrated planning across administrative boundaries and government scales that requires both formal and informal ways of cooperating with each other in order to gain positive results from large public projects. In the case of the THIP design, the limited perspective of its designers aggravated that point of taking various policy issues into account:

> "(…) when I asked engineers: 'Why did you not consider linking the uses along the highway (…)?' And they say: 'Oh, we were required to minimizing costs.' (…) Yeah, to them it is costs, you know. Not to serve those neighboring markets [and] centers. So it was not planned as a comprehensive development of a corridor. It was planned as a kind of thoroughfare to deliver traffic to and from the city (…)."
> (interview university researcher B)

> "I think the majority of people who were involved in this are people who are just highway engineers, who have never done transportation in the context of looking at traffic engineering and management, what are they calling it, ah, transportation planning per se. I think they were involved very little." (interview university researcher C)

A point that was made in the interviews as well as in other media (such as the report by KARA & CSUD 2012 or Klopp 2012a) is that the process of designing the THIP and other large transportation infrastructure projects in Kenya in general needs to be opened up. There is a limited understanding of who actually is a 'stakeholder' and of who can provide valuable input into a seemingly apolitical, technical infrastructure project. One interviewee described how features relevant to people along this transport corridor were not 'visible' to the THIP designers who solely focused on upgrading the road (interview university researcher B).[39]

Partly in contrast to this viewpoint, others interviewees saw the root of the THIP design process not in a limited perspective of executing project managers, but more ingrained in the Kenyan system of politics and planning related to the aspect of the historical impunity and corruption (cf. chapter 3.1):

> "(…) these institutions (…) most of the time are being run by politicians. And, sad to say, most of the politicians go there to see what they can get from there, apart from their salaries. Other than their salaries, what else can they get from there? So you get institutions being run down." (interview journalist C)

Since the Kenyan system is still very much characterized by the inherited British government system, the centralized, top-down structure dis-empowers lower-level,

---

[39] One illustrative story was mentioned during the discussion forum on the THIP (held 20 November 2012 at the University of Nairobi): The concern was raised that even the installation of footbridges after the completion of Thika Highway did not take into account that the Maasai people with their goats and cows in that area would not be able to cross the highway by using the bridge to reach the slaughterhouse, as these animals have not been seen to actually walk up and down stairs (Earth Institute 2012).

lower-rank bureaucrats, with the overall power resting with very few high-level decision-makers:

> "But when you come to the political, or higher-level decision-making level, the corruption and self-interest is so strong that even the lower technical cadres don't have a chance to impact on decisions. That is why you can have wonderful meetings, let's say, with the working class of ministries. You get really people who you think they have a genuine interest to improve things. But of course, if you don't have high-level decisions that enable them to do so, then there's not much they can do." (interview UN-Habitat expert A)

This comment also reflects the author's experience when talking to one local government bureaucrat who expressed a certain powerlessness towards higher-level government. In this interview, the bureaucrat made clear that the local government could have provided valuable input into the THIP and how it could best link up with existing plans and policies; however, these local authorities were never meaningfully engaged with the design and implementation of the THIP.

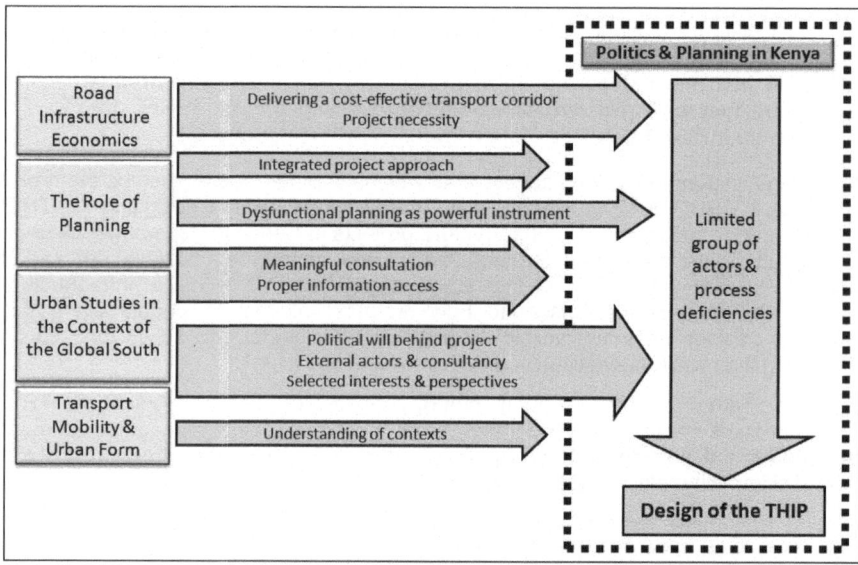

Figure 5.2: The design of the THIP vis-à-vis relevant theoretical perspectives and their critical themes, embedded in the politics and planning context of Kenya (Source: Author).

In conclusion then, the view on actors and the process of the THIP uncovers that a limited group of people decided over and planned this project and did or could not consider the heterogeneity of ideas, concerns, or recommendations that could have been provided by other stakeholders, experts, or the public in general. Likewise, the process was limited as people were not meaningfully engaged and even the most basic information of the THIP design was not (easily) accessible. One can see political aspects underlying these processes, as well as the overall top-down planning system with a narrow engineers' focus impacting on the project's design. Also aspects from the planning and the transport mobility/urban form literature came up in this discussion, when matters of project integration between urban development and infrastructure provision are concerned and context-sensitive design solutions for existing settlements would have been necessary. These aspects are shown in figure 5.2 with regard to the theoretical perspectives' input into the analysis of THIP's design in consideration of politics and planning in Kenya.

With the scope and scale of such a project already posing challenges of coordinated and integrated planning that were not fulfilled in the THIP design process, the empirical analysis of this project's implementation will illustrate how these process deficiencies resulted in path-dependent repercussions in realizing the THIP.

5.3 The Implementation of the THIP

> *"(...) are there conflicts of land uses? Yes, of course! This is Kenya."*
> *(interview local government bureaucrat)*

When discussing the implementation of the THIP, this analysis turns to the question of how the highway was actually constructed and what other measures were necessary to implement this project. This said, implementation features in this part of the empirical chapter are not to be confused with the THIP's outcomes – the identified changes refer solely to construction impacts on the project area and affected people *during* the implementation phase.

In summary, the construction period from the end of 2008 until the end of 2012 saw the wide expansion of the former four-lane road into an eight-lane highway accompanied in several sections with two service lanes on each side.[40] It is clear from the spatial scale of the construction that the impacts felt during this period were intense.[41] This concerned both people who lived and people who worked along this transport corridor as local residents, commuters, street sellers, shop employees, etc. As newspaper articles exemplify (see for instance Cheboi 2008a, b), the initial skepticism if the road would actually be built or if the project would become another 'white elephant' in Kenya's younger history of big political promises with little delivery, was quickly changed when bulldozers moved in and started to demolish shops, homes, and other constructions along the road. As the following quote from an interview illustrates, people really saw that their government was serious about implementing the THIP as planned when even a store of the country's largest supermarket chain, Nakumatt, was demolished:

> "(...) But given the impunity that we had in this country, I felt that it was time when someone had to do something (...) You know, (...), we needed a lesson. (...) And for me, I think, it was very useful that it was not happening to a small time guy. You know, it was happening to a big player! And that you were then seeing that it can happen to Nakumatt, then it can happen to anybody else." (local government consultant)

---

[40] The photos from various media sources that can be found in Annex M illustrate the vast spatial expansion during the construction and provide impressions of the implementation phase.

[41] The online "Thika Road Blog" (related to "MegaProjects Kenya") covered the implementation process extensively and fulfilled a valuable external watchdog role: http://thikaroad.blogspot.de/ (MegaProjects Kenya 2012).

Eventually, the project took one year longer than planned and the budget went up from 24 to 31 billion Kenyan Shillings. As was explained by two interviewees (local government consultant, professional body representative), the extent of cost increases and project delays were perceived as relatively acceptable for Kenyan standards and are also not uncommon for large transportation infrastructure projects in general.[42] However, it was again not the problem of delays and cost increases that sparked criticism, but the lack of proper communication about these issues. The THIP lacked an on-going communication strategy. Furthermore, during the construction, the traffic management by the executing Chinese companies was sub-standard in that the necessary diversion of roads and changes to the traffic system were chaotic and not sufficiently displayed. This resulted in congestion, accidents, and a number of deaths that could have been mitigated if international standards would have been applied by the executing companies (Oduor 2010; Gisesa 2012).

Photo 5.2: Road accident on Thika Road during implementation phase (Source: Suleiman Mbatiah 2009, http://www.flickr.com/photos/31513264@N08/4480917199/).

With regard to the THIP affecting existing businesses and settlements, the Kenyan Government dealt in a proper way with those people that had formal land titles. Going through a legally regulated process, these land owners were compensated for the land that was needed for the highway expansion. As the land experts amongst the interviewees explained, these land owners were very willing to sell their land, not only because the government had the legal tool to compulsory acquire that land for public uses (i.e. the THIP) as a last resort option, but also because good – some claim even above-market rate – compensation was paid (interview university researcher C, D). In this regard, Kenyans seem to agree that land has to be taken to accommodate for the infrastructure that is needed for the growth of their urban agglomerations.

---

[42] Reasons for cost increases were said to be rising fuel prices and increased construction material costs (Mathenge 2008; Onyango 2011).

In contrast to the acquisition of formally owned land, the government's dealing with informal land uses along the Thika Highway was significantly different (Cheboi 2008a, b): The short story would be that bulldozers moved in – with only short prior notice – and demolished residential and business structures that had been placed illegally on the strip to each side of the Thika Highway that had been demarcated as a road reserve in the 1970s. It is clear that this short story is too simple when the livelihood of people is concerned. The interviewees mentioned various aspects that play a role in this regard. First of all, many infrastructure projects that were planned in the 1970s had never been built and led people to wrongfully believe that the empty space next to roads is free or at least safe to be used. As one interviewee argued, most of the people that illegally use this kind of land have set up camp there due to a lack of other options (interview journalist B). This is a question of the government securing accessibility and affordability of land and housing to its people. On the other hand, interviewees were also right in connecting this 'land squatting' to the aspect of impunity in Kenya:

> "You know, they see this large strip next to the road being empty. And they think: Why is this plot not taken on such a prime location? Then they are setting something up there and then their friends will follow, and soon the whole road reserve is occupied. They are grabbing the land and first cultivate it a bit, because they know that it is not legal. Then they put up a small stand. After a couple of months, they build a small shack. After some years, when still no one came to demolish their stand and to take the land, they will invest in bricks for their buildings." (interview university researcher E)

> "I think that was back in history. There is a term which has become very, very common in Kenya now – called: impunity. I know something is wrong, but I just do it. Yes. I just do it. Because other people have done it and it probably worked out well with them, they made money, they had property, and so on. Why should I not do it?! So it's just, ah, I would say it's just a disobedience to the rule of law ...and it has been historical. If people did that at some point and no action was taken...somebody will do it and no action is taken, somebody else would want to do it, yes!" (interview university researcher C)

In addition to this informal type of land uses, there are other cases where people had faked land titles for parcels on the road reserve and had sold it to unknowledgeable people, who were in need of land. These buyers were made to believe that they had acquired a rightful land title, just to discover years later that they were tricked. These people then became victims without having deliberately violated the law. The land register system in Kenya is also not of help in this regard and the issue of fraud and/or conflicting land titles remains unsolved. Given this situation, the underlying question is if or how the government can deal with proper justice in relation to illegal squatting of land that is born out of multifaceted reasons related to need for land, impunity towards misuse, and faked/conflicting land titles. One bureaucrat described the government's role as follows:

> "You know, if you commit a crime or if you do something that is illegal, it is not up to the authorities to know where you go. (...) So I think there is a misplaced notion that we have to take care of people who have disobeyed the law. I think it is misplaced. I don't know what

they do in your country when people disobey the law...do we find another place for them? The only place we find for people who break rules, we used to take them to court. (…) It is deliberate and people should learn that when we reserve highways, they should respect that the highways must build up for." (interview national government bureaucrat B)

In a similar vein, another interviewee argued:

"(…) Because it was illegal. You know, (…) in any part of the world, if you compensate somebody for illegality, it will increase the illegality. (…) I will now in terms of wanting to be compensated go and put my property somewhere, so that when they now want to use it, they will pay me back. But when it becomes punitive, then it is good, because it discourages people from going into property which does not belong to them." (interview university researcher C)

Both interviewees' comments are legitimate with regard to impunity having characterized Kenya's history of land uses for long and the government needed to show a serious stance on this matter when implementing the THIP. However, there is still one aspect that was missing in many interviewees' responses to that topic: If the government decides to teach people a lesson that the road reserve is not to be used illegally and that people will lose their property when they build on it, the government still has to decide on *how* to actually teach this lesson. In the case of the THIP, there was more damage done than necessary – both with regard to residential and commercial land uses (Ombuor 2012a, b; interview university researcher B, civil society organization representative, journalist B). During the construction phase, businesses were blocked off from their customers, existing informal market places were destroyed, and, therefore, the livelihood of people who depended on these small business activities was put at risk without due consideration by the executing authorities and companies (Muiruri 2012, Cheboi 2008a, b). This did not happen because the highway was built. It happened because the government did not care about assessing and then mitigating the impacts one could expect from a large-scale project like the THIP:

"(…) I think the approach of doing that was an issue. (…) most people [were] never given notice, they were never informed about what the plans were. Somebody woke up one morning and discovered that their shed had been demolished, the road kiosk has been demolished. So they were not given proper notice to be able to take the right action. (…) we needed to get to a point where we're removing this person from the road reserve, but again how it was done was a pity (…) and it didn't have proper mechanisms of giving the full notice and giving them alternatives. Because if you remove somebody from one place, maybe that person is on the road reserve, but I think as a Government it is a responsibility measure that you think of alternatives – where they are going to go to." (interview civil society organization representative)

What one investor in an interview metaphorically described as "You can't make omelets without breaking eggs" would be an over-simplified conclusion here, since decision-makers still had a chance of deciding on 'how to break the eggs' and what options to offer to affected people. Therefore, this part of the empirical analysis, which is summarized in figure 5.3, exemplifies how a macro-level project needs to be analyzed (and ideally also planned) in consideration of its micro-level impacts. The implementation of the THIP brought with it necessities of expanding the road, taking land, and destroying im-/ material values. There is no ground of disputing this aspect. Nevertheless, the style of implementing necessary tasks and anticipating repercussions tells something about how these kinds of projects are done in Kenya and how decision-makers perceive *informal* uses as *illegal* uses for which no careful consideration is required. While compensation was handled smoothly in the case of formal land owners, informal land users (both residential and commercial) were declared to have acted illegally and thus were denied any right to receive proper mitigation matters. Decision-makers needed to deal with the history of impunity in Kenya and the THIP was an option to teach a lesson, but this does not relieve the government from its responsibility to care for all its citizens and their livelihood/survival. Businesses had to be destroyed physically and/or economically during the THIP construction, but the damage was much greater than if the plans for the THIP would have included mitigation matters and their actual implementation.

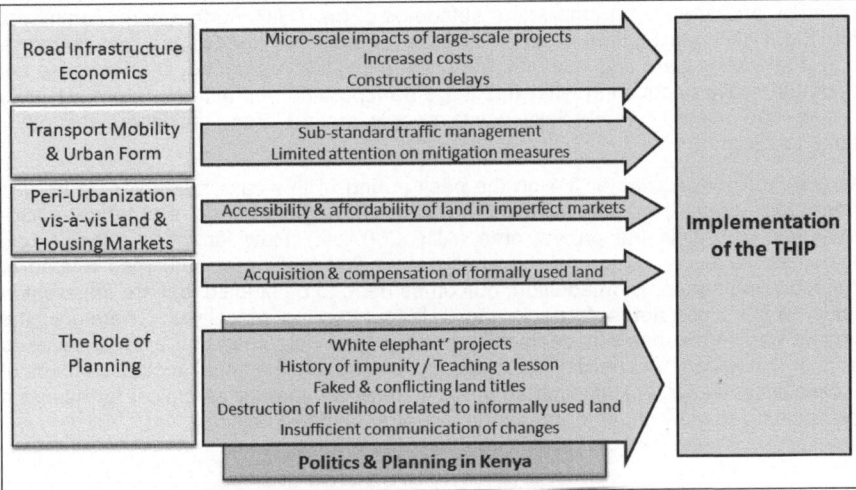

Figure 5.3: The implementation of the THIP vis-à-vis relevant theoretical perspectives and their critical themes, embedded in the politics and planning context of Kenya (Source: Author).

5.4 The Outcomes of the THIP

"Losers? (...) No. I think it benefits everybody." (interview investor)

After having discussed the idea, design, and implementation of the THIP, one can now turn to the THIP's outcomes. Repeating the remark from chapter 4.1, this thesis does not aim at naming every single outcome of the THIP, since a large transportation infrastructure project is always a matter of process and one cannot foresee external changes that might influence how outcomes of the THIP play out in the medium and long term. Therefore, this part of the empirical analysis presents outcomes that are already apparent in the study area and those outcomes that are very likely based on how this project had been designed and implemented.

Analyzing and presenting these outcomes is no easy task, since the THIP has been a project of such a broad scale and scope that various thematic areas are affected. A generally established view on the outcomes of large infrastructure projects is that of 'winners' and 'losers'. However, this viewpoint is very often limited to a focus on a project's beneficiaries, while its losers are disregarded. When specifically mentioned winners and losers were coded in the interviews, it became clear that there is a huge heterogeneity in identified stakeholder groups benefitting from the THIP. For interviewees it was easier to assign benefits to specific groups, while the description of negative outcomes and how they particularly affect certain stakeholders seemed to pose difficulties. As one interviewee correctly highlighted (interview university researcher A), this benefit-bias is not helpful in understanding a project's repercussions and, in a second step, mitigation measures that could have been taken, both in the design and implementation stages as well as afterwards.

For that reason, general conclusions such as "Thika Highway foremostly benefits Kenya as a nation" (interview university researcher E), or "There is no one who is not a beneficiary" (interview investor), are too broad to seriously help to assess positive, negative, and often even ambivalent outcomes of the THIP. Furthermore, comments that the THIP was thought to address broader, national needs, and that therefore individual needs were less relevant (interview university researcher D[43]), are also not helpful in really uncovering what this large transportation infrastructure project does to the road users, commercially active people in that area, residents, and other stakeholder groups.

As was outlined in chapter 3.4 on the background of this case study, the THIP was meant to improve the traffic situation, facilitate interregional trade, and trigger socio-economic growth in the project area (ADF 2007: vi). How far these rather broad objectives have played out will be analyzed by first looking at outcomes related to transport and economy. In addition, outcomes need to be studied that are apparent in fields which were not a focus in the THIP's intended objectives. Therefore, the second part of this analysis deals with those project outcomes that broadly relate to land and housing. A specific focus will be placed on the multifaceted concept of accessibility, which was highlighted in the theoretical chapter as critical for having a (re-) politicized view on large transportation infrastructure projects.

---

[43] This interviewee's conclusion was: "(...) the needs are national needs, more than individual needs. And ordinarily (...) we look at the benefits, nevertheless, when we are going to describe between right and wrong, or what you should do and what you do not do. If you look at it in terms (...) of the results, then we're looking at the greater benefit to the greater majority. (...) what else the individual wants, or the minority we're talking about. Essentially when you're doing a development project, we won't satisfy everybody who has to do with it" (interview university researcher D).

5.4.1 Changes Related to Transport and Economy

Photo 5.3: Informal market area at a central matatu stop around Ruiru (Source: Author).

Traffic, Transport, Trade

Since the opening of the new Thika Highway in November 2012, the travel time between Thika and Nairobi has decreased significantly. Even though everyone has a different time-keeping, the average estimate (from field visits, interviews, and conversations with locals) is that it is now three to four times quicker to move between these two urban centers. Furthermore, the predictability of how much time someone will actually need to travel from A to B has increased. Less travel time has already translated directly into cost-savings, since public transit fares have been decreased and individual road users can save on gas and wearing down parts of their cars (brakes, tires, etc.) that were formerly more stressed due to heavy traffic congestion. The economic relevance of the time saving is exemplified by one pharmacist who works in Juja:

> "You know, the big benefit for my business with regard to the commuting time is that I can actually work longer – I can have my pharmacy open from the early morning to the evening. I do not need to close down earlier anymore because of the heavy traffic that will necessitate long time to get home. Now I open longer and get home quicker. Either I go by matatu or by my own car." (interview local pharmacist)

These positive outcomes appear somewhat 'incomplete' when several other aspects are accounted for: As was already criticized, besides a decrease in matatu fares,

71

transit improvements (particularly important for lower and middle-income people; cf. Salon & Aligula 2012: 69) were not realized, since the bus rapid transit system that was planned for the THIP has not been developed. Non-motorized transport uses, despite their important role in Kenya (ibid.: 69), have also not been taken into account sufficiently (as observed and tested by the author in the field). Even though footbridges were introduced in reaction to public debates on how to actually cross an eight-lane highway, there are still very few.[44] The author has made a self-experiment and walked along the highway between Ruiru and Juja, as well as Juja and Thika, and the distance between crossing options is sometimes as far as two kilometers. This is a distance that causes people to cross the highway just where they are, resulting in dangerous maneuvers that put themselves and other road users at risk. The THIP and its positive outcomes with respect to transport are thus biased towards motorized transport, and even more so towards individual car use.

In addition, several interviewees criticized that this highly improved road is not (yet) supported by a functioning road system: some access and feeder roads have been provided, but many more are still needed to really open up peri-urban areas along that transport corridor. In the case of Nairobi, it is rather a matter of traffic management, since the Thika Highway now ends directly in the city where the existing traffic congestion issues have still not been addressed (see photo 5.12). As a result, commuters – as several of the interviewees are – might need 30 minutes instead of two hours to go from Thika to Nairobi, just to add another 30 minutes for the last four kilometers in the central part of the capital city. This clearly highlights the importance of integrative transport system planning and already hints at the challenge of politics and planning across administrative boundaries. Furthermore, the Thika Highway requires proper, well-established maintenance to be sustained, as it is a road of elaborated standards in comparison to most other roads in the country. As interviewees and newspaper articles (for instance Opukah 2012) indicate, there are well-founded concerns that the road quality will quickly deteriorate if a so-called 'maintenance regime' is not developed where responsibilities for the road maintenance are clearly assigned and sufficiently supported by human and financial resources.[45]

The intended objective of facilitating inter-regional trade – also with regard to the African Development Bank's vision of transnational transport corridors – is rather difficult to assess in terms of factual outcomes. The highway improvement did address the 'bottleneck' problem of the former road with its low capacity. But in order to live up to its broader-scale objective, the THIP only makes sense together with other road projects that are currently under construction, such as the envisaged LAPSSET transport corridor, and specifically the extended Thika Highway up to Isiolo and Moyale. Once these projects are realized, inter-regional trade can increase – at

---

[44] Although one interviewee underscored that in the more urban parts of the Thika Highway the provided footbridges have increased the safety particularly for school children to cross the street (interview journalist B).

[45] This aspect becomes even more urgent in a situation where metal parts of the Thika Highway road furniture have been stolen since the implementation phase (Onyango 2012; Sangira 2012). For the financing of the maintenance of Thika Highway, the introduction of toll stations has been discussed since 2012 (Omondi 2012; Kihanya 2013). While it is still open to debate if a toll road would be legally possible, it would provide significant funds for maintenance, although to the detriment of lower-income people who would be hard-hit by this extra transportation cost, which would also contradict the original objective of making mobility along this transport corridor more affordable. Also with private car users this is not very popular, as 9 Kenyan Shillings per liter of fuel already go into road maintenance.

least based on the criterion of physical road infrastructure features (not regarding important questions of trade policies and border customs controls).[46] This assumption, however, was questioned in an interview by one investor who has a large-scale project along Thika Highway illustrated the differing viewpoints of transnational trade linkages imagined by the government and international actors on the one side and investment criteria of private actors on the other side:

> "Investor: No, we are certainly not making investment decisions based on these broader international pieces of infrastructure.
>
> Interviewer: Because your investment decisions have a smaller time frame?
>
> Investor: Yes. (…), we have to enter and exit our investments within a five to ten year period." (interview investor)

Photo 5.4: Food stands and boda boda at matatu stop between Thika and Juja (Source: Author).

Business and Job Opportunities

As the appraisal report by the African Development Fund (2007) before the project implementation argued, the THIP would open up the Northern Nairobi Metropolitan Region to investment that would trigger economic growth. Field visits, newspaper articles, and the interviews all confirm an increased vibrancy and dynamic in this area since the construction and final opening of the THIP (see for instance Wairimu

---

[46] See Annex C for corresponding maps showing THIP's linkage to other road infrastructure projects.

2012c). Big investments have already been confirmed (Iraki 2011; Kamau 2013), although some announcements need to be treated cautiously as they might prove not very lasting due to various (external) economic factors that might change a company's investment calculus (interview investor).

The peri-urban project area is believed to primarily experience investment in the manufacturing and service sectors, much less in industries.[47] These investments are market reactions to the increased number of middle-income people that are moving into the Northern Nairobi Metropolitan Region due to the opening up of the transport corridor and adjacent areas. Important to an assessment of new or increased business opportunities is also the design of the highway and its access points. As one interviewee explained:

> "There are cases where probably businesses will have to die. But there are cases where they will also have to come up. (...) The road has created certain nodes. And those nodes have brought up businesses. Where the road has removed the nodes, the businesses died. But where it has created them, particularly where you have the underpasses, you'll find most of those places businesses booming up again." (interview university researcher C)

This aspect refers to the underpasses that were constructed to link existing settlements and market places in cases where the design criteria for international trunk roads necessitated an elevation of the Thika Highway. While some businesses lost their advantageous position along the road, others have seen an improvement, as it was not only described by interviewees, but particularly by locals along the THIP corridor. Furthermore, new businesses can place their structures strategically at the new growth nodes. It only has to be asked what informal businesses gain from these changes: Due to the expansion of the road, informal businesses on the road reserve were demolished and had to move. Some of them were placed further inside the land, which meant that they were removed from the lively trading areas along the road, which has negatively impacted on their businesses (interview African Development Bank expert; also see Ngirachu 2012; Ombuor 2012a, b). On the other hand, most of the informal businesses were mobile (small kiosks) and thus moved to the new growth nodes and continue to enjoy close distance to their customers.

Informal businesses in competition to formal businesses, which now increasingly move into that area, can also be seen from the perspective of the customer. And here, two views come into play: While small, often informal businesses are closer to settlements and easily accessible (Becker 2011: 5), their prices are often higher than those of the big retailers, particularly with regard to supermarkets (interview UN-Habitat expert A, investor). Thus, small traders might be able to sustain their livelihood, but their prospective customers might benefit from increased affordability offered by the new formal businesses. As was discussed in chapter 3.2 on Nairobi, one should also critically assess the characteristics of these informal businesses, as one interviewee did:

---

[47] The interviewed investor pointed out that the skyrocketing land prices in the Northern Nairobi Metropolitan Region make it very difficult for industry to set up new facilities there, as they cannot compete with the intense and dense land uses by manufacturing and particularly service sectors (also see Chege 2012). Nevertheless, existing industries were said to remain in that area and a few new developments even in the industrial sector are likely to happen (interview university researcher B, C).

"This is what we call the jua kali or small-scale businesses in Kenya. And if you observe them well, they sit on huge stocks that are not sold. (…) So if you compare to what they sell and what they produce, I mean, there's a huge gap. Yeah, they are overpriced normally, because, the less you sell the more you try to get out of one piece you sell. So they're not really competitive compared to really big commercial outlets. They do have clients, but certainly not enough. And also, they don't have mass-production capacity. So if you go to such a guy and say: 'Look I have a school. And can you do all the desks?' Normally, they can't. (…) It's copy-cat business. These people do not go into innovations, because innovations are risky, and if you're poor you cannot run a risk. Hmm, so you do something your neighbors are also doing." (interview UN-Habitat expert A)

This raises the question of whether these informal business owners would not prefer to seek the opportunity to get a job in one of the many new businesses that are opening up in the Northern Nairobi Metropolitan Region. Here, interviewees disagreed with each other. Some stressed the entrepreneurial nature of Kenyans who want to do something on their own (for instance interview professional body representative). Other interviewees (for instance African Development Bank expert and journalist A, B, C) called attention to the previous economic situation along this transport corridor:

"(…) I mean, you can ask anybody of these small operators, roosting maze at the corner, or doing small carpentry works or whatever, what they would prefer: a proper job or doing this jua kali business. I guarantee you that 90 per cent would prefer the proper, salaried job, which gives them security. Because the jua kali business is a very insecure venture. Sometimes these people have a business, other times they don't have." (interview UN-Habitat expert A)

One does not need to find a final judgment on the informal businesses and the relation to the THIP and its outcomes. However, it can be summarized that there have been impacts during the implementation when businesses died. And also amongst the outcomes, there have been and will be cases where businesses have no future due to the new spatial features of this transport corridor with closed-off zones and new growth nodes (also see Muiruri 2012). On the positive side, as many interviewees confirmed, both for formal and informal businesses, the opportunities have increased in that area. Workers, skilled or unskilled, will have very good chances to get a job. But with regard to the informal businesses, one also has to conclude that the transformational process that was triggered by the THIP was not used by the government to provide demarcated spaces in the new growth nodes for informal businesses – also as a step to facilitate their integration and formalization at a later point. Such a formalization would not only offer these informal businesses greater security for making investments, but it would furthermore increase local authorities' tax base as they can only tax formally registered businesses. This has been a missed opportunity in the THIP.

Photo 5.5: Cattle drover with cows near Thika (Source: Author).

Agriculture

Even though services and manufacturing are gaining momentum in the Northern Nairobi Metropolitan Region and the country in general, the agricultural sector still sustains the livelihood of people along that transport corridor by providing farm produce, land for cattle, and jobs (see for instance Wairimu 2012a). These peri-urban areas with their long history of agricultural production are very important for the supply of farm produce to the metropolitan region.

The logic behind the THIP was to improve the mobility of goods and people, which would benefit farmers in the project area, since their land would be better accessible, they could more easily access market places, their transport costs would decrease, and thus the customers would save on food produce costs (interview African Development Bank expert, journalist A). The only miscalculation in this logic is that the opening up of the peri-urban areas made this land also open to different uses. While this topic will be analyzed with further detail in chapter 5.4.2, it can be concluded that the anticipated benefit of the THIP on the agricultural sector will not pay off: Similar to Calderon Cockburn's theory on urban and rural functions of land profits (1999: 5-8), land owners in the Northern Nairobi Metropolitan Region are taking their 'once in a lifetime' opportunity and subdivide their agricultural land and put it into different, urban uses (Standard 2012). As one interviewee explained:

> "We the professionals would say, you know, what we've land for agriculture will be land for agriculture. The owner would look at it and say: 'If I'm getting so much from my cows or coffee, but I can get five

times or ten times more, I actually now go for this [selling the land].'"
(professional body representative)

This well-known story is very common and depicted, for instance, in Thuo's study of rural land conversions in Kenya (2010). He quotes one farmer:

> "I was a coffee farmer. We used to get a lot money... All of a sudden prices became so bad, I was in debts. Then some people approached with an offer to sell a piece of land along the main road.... The offer was irresistible. With the sale, I was able to clear the loan and build some rent units myself... Further, I was able to pay school fees for my children. The income from the rent the tenants pay is more than what I used to get from coffee.... I just laugh at my neighbours who still grow that thing... Others are now 'seeing the light (Farmer #2)." (ibid.: 3)

In his study (ibid.: 3), Thuo makes clear that the money from selling land is actually not spent in a strategic and sustainable way, but most often for consumption. Therefore, one could argue whether farmers in peri-urban areas who sell their land are benefitting in the long-term from the THIP and its impact on changing rural to urban land uses. Furthermore, their land has to be seen in a broader context of food insecurity in Kenya (interview university researcher B, D, national government consultant; also see Wairimu 2012a). The area between Nairobi and Thika features very important arable land that is now being converted to non-agricultural uses and is most often lost irreversibly (ibid.; also see Chege 2012). Since the government has not planned for this rapid conversion of parcels and loss of agricultural land, problems in this regard are likely to increase:

> "That is my contention: Some of the agricultural land must be preserved. And, we have problem of food security. So it's not an easy-peasy project of small matter. You cannot just talk about roads, you cannot only talk about health, you cannot talk about that. Food security is a major national policy issue in this country. And how do we do it if all the agricultural land is lost?" (interview university researcher D)

The same interviewee also discussed the very feasible option of introducing commercial, high-produce farming in the THIP corridor area; however, the interviewed national government bureaucrats saw their country going in a different direction dis-/ missing the importance of food security. For them, agriculture in Kenya does not play a relevant role in the national economy and exports, since products (such as coffee and tea) are not processed to the extent necessary and do, therefore, not provide sufficient benefits. They want to follow what they describe as 'the Brazilian example' of industrializing their country at the risk of underestimating the importance of agriculture for the livelihood of their people. Arguments that the peri-urban areas between Nairobi and Thika feature mostly tea and coffee and that these two products are not food are narrow-minded, since they can nevertheless provide an income to farmers. Furthermore, there is also horticulture and dairy farming in that area, and if the government would support farmers, a change from, for instance, tea to high-value crops would be feasible (interview university researcher D).

Conclusion

What becomes clear from this part on agriculture is that the THIP's outcomes of opening up the peri-urban areas to investments have resulted in a loss of agricultural land that is connected to issues of food security, government support for agriculture, and theories on rural and urban land use profits. It shows that large transportation infrastructure projects need to be planned, ideally, in an integrated fashion with other policy areas, or at least in consideration of them. The infrastructure construction-economic growth causality that is so often employed in these projects will not be realized if the road project is done in isolation.

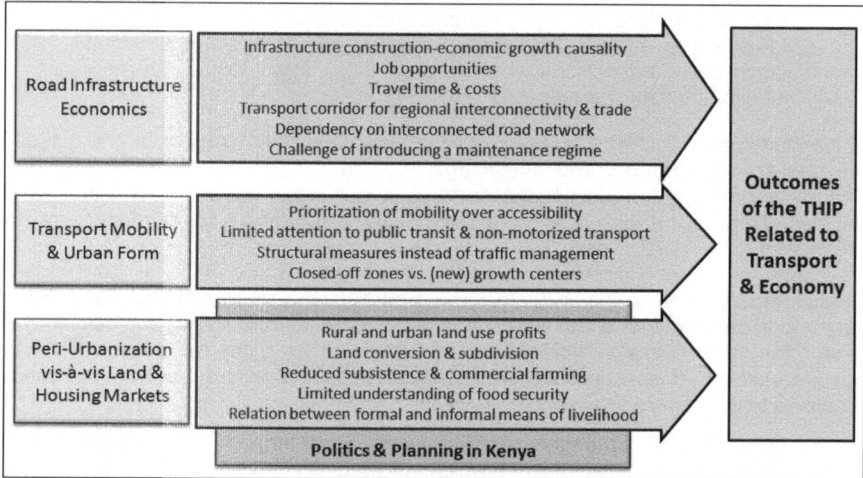

Figure 5.4: The outcomes of the THIP related to transport and economy vis-à-vis relevant theoretical perspectives and their critical themes, embedded in the politics and planning context of Kenya (Source: Author).

There are economic benefits from the upgrade of Thika Highway in reaction to pressing traffic congestion. Thus, one part of the causality holds true in that economic development (new business opportunities and jobs) follows an infrastructure investment. However, some negative outcomes could have been mitigated if they had been assessed prior to the project implementation. The illustration of outcomes in the agricultural field shows that intended benefits can be reversed when a better road leads to improved mobility and accessibility, but initial cost savings are balanced out again by the increased travel costs for farm produce that has now to be brought in from farther away. All these aspects are again summarized in figure 5.4, showing the relation to politics and planning in Kenya and the relevant theoretical perspectives of which some critical themes came up in this discussion.

A few interviewees also raised an important concern with regard to the infrastructure construction-economic growth logic: Do the outcomes of the THIP prove that the project as it was done was necessary to achieve this kind of development? Or to ask differently: How do we know if only six lanes (instead of eight) and an improved traffic management would not have resulted in the same outcomes? The idea behind these questions is to acknowledge the need for transportation improvement in that corridor, but to challenge the assumption that a 'superhighway' somehow automatically brings with it economic investments:

> "(…) there is sort of an abstract notion of economic development that will kind of emerge of having that kind of infrastructure (…). However, I think it doesn't really reflect cutting-edge ideas about how we really should build infrastructures, especially infrastructure in densely populated areas. So this tragedy that it is locked into a modernist vision that is actually not cutting-edge and that is highly critiqued in other parts of the world. (…) So, I don't think that we didn't need to build a bad, like poorly designed superhighway in order to encourage commercial businesses and linkages to the smaller businesses." (interview university researcher A)

While the discussion in chapter 2.1 has already shown that research findings are ambivalent in that regard, one can deal with this question by studying the repercussions of the THIP more closely in order to evaluate the relation between positive and negative outcomes. For that the following analysis of outcomes related to land and housing is revealing, since it sheds light on thematic fields that were not a major focus in the project's original objectives.

5.4.2 Changes Related to Land and Housing

> *"In Kenya (…), it all rests with the private market. You get it?!" (interview national government bureaucrat B)*

Photo 5.6: Billboard at town entrance to Thika advertising real estate project "Buffalo Hills" (Source: Author).

Land-Use Changes

In order to understand socio-economic changes to land and housing in the THIP corridor area, the actual spatial changes need to be described first. These changes correspond with descriptions that were already presented in chapter 2.3 with regard to peri-urbanization literature: The peri-urban Northern Nairobi Metropolitan Region – as confirmed by all interviewees and depicted in the media – has been experiencing a rapid change to its land (uses) since the implementation of the THIP. As illustrated in the photos in Annex D, the peri-urban area is starting to be filled with new housing developments. This all happens in a very scattered way, with no apparent structured development pattern (field visit notes; also see Ndoria 2011: 28-29). Single homes, apartment buildings, or other types of housing are popping up everywhere in the landscape, and sometimes it is hard to figure out how these parcels are accessible from the highway and its feeder roads or are provided with other infrastructure and basic services (see next section below). Nearly all interviewees were mentioning the reasons for this kind of development: First, land owners, investors, and developers are seeking the profitable opportunity of quickly putting land into residential and commercial uses in a situation of land opening and heightened dynamics resulting from the THIP. Second, there are no proper plans in place to somehow control or even guide these developments. One interviewee described how there is a difference between giving private actors sufficient freedom to make investment decisions on the one hand, and ensuring the functionality and balance of land uses through development control on the other hand – particularly when congestion and fragmentation follow from the land development, it will impinge upon every stakeholder (interview real estate agent A).

For that reason, one interviewee pointedly termed the expected changes to the project area as "mixed development by default" (interview UN-Habitat expert A). And the interviewee was not thinking about mixed uses as a positive feature in this case, because this development will pose a risk to the ecological and also political-organizational sustainability of the peri-urban area along that corridor. While the former aspect of ecological sustainability relates to questions of ecosystem services in relation to the provision of infrastructure and basic services (see below), the latter aspect of political-organizational sustainability deals with questions of how government can actually manage and cater for these fragmented areas in order to ensure an equitable long-term development of peri-urban settlements.

On a descriptive level, one can observe that existing towns, such as Ruiru, Juja, and Thika, are developing into vibrant growth centers along the Thika Highway corridor. Other new settlements will also emerge at the nodes where the highway connects with feeder roads. What is to be underscored here is the tremendous pace with which these changes happen and the symbolic interlinkages that develop between urban and peri-urban areas in the Nairobi Metropolitan Region (see also Mwongela 2010b):

> "You know, five years ago, places like Juja were not even thought of as residential areas...We only had villages there. People who were born around there were having their homes there. So it was, what (…) we call it 'mji mdogo'. Is like a village [English translation: small town]. It used to be like a village. So, it was hard to think that somebody was living in Nairobi, while finally decide to go and rent a house there. We didn't have rental houses there. But because of the opening up of the area by the coming of the superhighway, now developers are seeing that area as a potential real estate growth area. And they're developing." (interview journalist C)

It is this aspect of real estate development and opening up of formerly 'sleeping' areas that this analysis now turns to; first by discussing how infrastructure is provided in the peri-urban area and then by illustrating what kind of new land and housing developments are happening in the Northern Nairobi Metropolitan Region.

Photo 5.7: Stand-alone apartment development on the greenfield between Thika and Juja (Source: Author).

Infrastructure and Basic Services

In light of the tremendous development of land along the THIP corridor, the provision of infrastructure and basic services[48] is key for a functioning use of residential or commercial land uses. Some interviewees – particularly those closest to government – downplayed this challenge and imagined that the infrastructure would be provided step-by-step following the gradual influx of new people into the area. However, those interviewees who have been studying closely the changes over the past years argued that the influx of people has already led to much higher pressure on the existing infrastructure system – to an extent that people in that area have been experiencing water shortages and power black-outs for several days in a row (interview journalist B; also see Wairimu 2012b). Infrastructure and basic services are currently sub-standard in these peri-urban areas and the challenge of catering for the residents' needs in the future will severely aggravate (Njenga 2013). In that regard, interviewees expect the provision of infrastructure to happen only partly, most likely in the growth centers, while other parts will remain cut-off from the trunk infrastructure system.

---

[48] See footnote 10 for definitions of infrastructure, basic services, social services, and ecosystem services.

In addition to this challenging situation, the area is also experiencing illegal uses of formally provided basic services. As has been proven in many neighborhoods in Nairobi, residents will find their way to connect with the public utilities system – no matter if they have to turn to formal or informal means (interview university researcher A). The problem here is that the already stressed system is put under even greater pressure. Given the fact that the THIP was meant to open up the area to residential and commercial uses, the Kenyan Government could have not really been surprised to see the THIP corridor losing its balance with regard to basic service provision; although some interviewees thought that this was actually the case.[49]

While the opening-up objective was recommendable, the lack of integrating the road infrastructure planning with the municipalities' and public utility providers' provision plans has already and will increasingly have an inhibiting effect on other positive outcomes of the THIP. Several interviewees described how the government has given up addressing these challenges in a pro-active manner. Instead – and this is also apparent in the THIP corridor area – government is witnessing, for instance, the fragmentation of land and the scattered development of housing, before it reacts to insufficient infrastructure provision. There is an underlying issue here with inter-governmental relations, since the aspect of infrastructure provision is another example of design and implementation flaws of the THIP that resulted in path-dependent repercussions that are rooted in a lack of coordination and cooperation between national, county, and local government beyond administrative boundaries.

Shifting the focus from the public to the private sector, theory would still claim that private actors could take over the public task of providing infrastructure and basic services (Dowall 2003: 19). In the case of the THIP corridor, this is only partly the case. On the positive side, the ability of the private sector to take over the provision tasks was confirmed in the interviews – particularly by the investor and real-estate agents, even though they criticized that no government incentives exist to actively deal with this task. However, there is a limit with regard to scale and scope to which developers can provide infrastructure. For individual parcels, the task is often not a problem and normally included in a developers' work portfolio of opening up land for housing construction. For a whole estate, developers can also provide larger infrastructure systems. In these cases, a clear differentiation has to be made: In upper-market real estates the infrastructure and basic services are not only provided, but even green energy and other sustainable solutions are included and used as selling points (interview real estate agent A, B, UN-Habitat expert A). On the other hand, lower-income estates are hardly ever equipped with sufficient infrastructure (interview local government consultant, professional body representative; also see Njau 2010: 54; UN-Habitat 2012a: 6, 9, 11):

---

[49] National government bureaucrat A provided a critical reflection on how the Kenyan Government has been dealing with this issue: "(...) very few people leave serious thoughts to the many impacts the highway will have. Maybe we can anticipate certain things (...). But they never anticipate other things like we need to increase the infrastructure and services to delivery levels, because populations are going to probably go up, and so on and so forth. And for many years in this country I don't think we approached it that way. We saw the highway as one off, which was taking already so much of our money, so we're not thinking about anything else. And when the problems present themselves, we start dealing with them. That's the way we traditionally looked at it. But now, people must begin to think about the whole range of impacts, including (...) ribbon urban development that is happen to take place along highways. (...) You're going to have development without infrastructure and services. And that is a new phenomenon for us. We're just beginning to think about how to respond."

"What we have been seeing is that developers are building their own road (...). (...) when it comes to a controlled development, where maybe a gated community, that's an example where people are coming together and developing their piece of road. (...) Yeah, a gated community is better, because it's well maintained and this is better. If you control that, if they would get the neighbors to come together. Others then where maybe each one on its own." (interview real estate agent A)

Beyond this difference between upper-market and lower-market estates, private developers do nevertheless reach their capacity and capability limits when trunk infrastructure is concerned. The interconnecting public utility lines and the overall access points for water, electricity, roads, etc., cannot be provided by individual actors in the private sector (interview investor, real estate agent A, B, UN-Habitat expert A). Even more important: Without the trunk infrastructure, the private sector activities on individual parcels can hardly function – they might use on-site diesel generators for energy and dig a hole into the ground to access groundwater, but it goes without saying that these 'solutions' are not long-term and/or effective enough to provide for the functionality of a whole peri-urban area that stretches from Nairobi to Thika, covering more than 2,500 square kilometers.

In addition to this aspect, there is also the problem of an over-reliance on a market logic that assumes necessary private sector investments in a strategic way. However, private actors – in theory and practice – can only make informed decisions, if information and guidance by the public sector is provided (Dowall 2003: 18; Grant 2009: 142; Rakodi 1997c). With the THIP being designed and implemented as an isolated, non-integrated project and the government not putting enduring support and funding behind the opening and development of the Northern Nairobi Metropolitan Region, private actors can only guess where infrastructure will be provided, what areas will become new growth nodes, and where investments will pay off. Besides, it has to be pointed out that private investors do not make these investments on their own costs. Instead the costs for providing such infrastructure is allocated to the final selling price, thus making land and housing more expensive (interview real estate agent A; also see Njiru 2013). Therefore, land development and infrastructure provision further impact on the land and housing markets in the THIP area.

Photo 5.8: Real estate agent signs along service road of Thika Highway between Juja and Ruiru (Source: Author).

Land and Housing Markets

The outcomes of the THIP with regard to the land and housing markets might be the most important and fundamental ones. These markets have significantly changed since 2009. A former less marketable peri-urban area along one of the most congested roads in the country became one of the prime investment corridors in the Nairobi Metropolitan Region (Okoth 2012; Wahome 2013). Estimates given by many interviewees can be confirmed by studies from the Institute of Surveyors of Kenya (Kanyangi 2010; Njau 2010; Ndoria 2011): land and housing prices in the THIP area have increased tremendously. Depending on the particular location of a parcel, prices went up between five to ten times in only a couple of years! This means that a parcel that has been sold for two million Kenyan Shillings now costs between 10 to 20 million Kenyan Shillings (i.e. from 11,600 to 116,000/232,000 US-Dollars).

Why have the prices skyrocketed? As was already explained, the upgrade of Thika Highway made the whole area more accessible through improved transport conditions. The announcement of the road construction caught the attention of investors (Kumba 2010; Ayodo 2012). What happened in the THIP area and what is common in Kenya is:

> "(…) when you build a nice good road, the moment people get wind of this – and this is of course long before –, those who have a lot of money will make sure that they will get prime plots along the road – still at relatively cheap prices to make a killing later on by developing land." (interview UN-Habitat expert A)

At the same time, the opening up of the peri-urban areas has to be seen in relation to a general land and housing shortage in the Nairobi Metropolitan Region (interview journalist C; Iraki 2011). Therefore, land and housing owners are speculating in the project area, but this speculation is in line with the market situation of supply and demand:

> "Kenya is a textbook capitalist market economy. Housing prices resemble the situation on the market. Right now demand for plots along the highway is high. If the offers are too expensive and no one is able to afford them, the prices will go done. People will not hold on to their land and keep it vacant. They want to sell." (interview university researcher E)

> "Because Kenya is an open economy, it's where supply and demand dictates everything – it's actually a capitalist economy. So at the end of the day, it's a willing buyer, a willing seller." (interview civil society organization representative)

Here, the road infrastructure theory on highway construction and intra-regional urban development is instrumental (cf. Boarnet 1998), because the above-described relation of supply and demand is not only connected to changes along the Thika Highway, but to the situation elsewhere in the Nairobi Metropolitan Region:

> "And that areas are also competing with other areas that are opening up. (…) The growth momentum can continue. It will continue to go on, up to some point, when people feel this one is way beyond as what we do. We look for alternative places and we're just moving on. Some years down the line, many other places will also be coming up. Yeah, once we continue with the infrastructure upgrade." (interview journalist C)

Since the infrastructure provision is a critical issue for any area to be developed, this interrelation of basic services provided and land and housing developed for the market will influence how areas change. In the case of the Northern Nairobi Metropolitan Region, a former "non-profit area" (interview journalist B) became a hot-spot of real estate investment (Wahome 2011). As was indicated earlier, original land owners use this 'once-in-a-lifetime' opportunity to sell their land, often without knowing where to go and what to do without their land in case they did not retain a small subdivided part of their parcel for themselves (interview university researcher C). This situation needs to be looked at in detail with regard to both ends of the markets: changes to affordable as well as upper-market housing.

Photo 5.9: Construction of middle-income apartment blocks between Nairobi and Ruiru (Source: Author).

Affordable Housing

Concerning the huge demand for land and housing as well as the accelerated development in the THIP area, a look at the new real estates along that corridor illustrates that only a specific supply and demand meet, i.e. upper-market housing (Wairimu 2012c). None of the interviewees see the peri-urban area along Thika Highway to develop into something different than a segregated zone where prime real estate will exist next to informal settlements where those people live who service higher-income residents of the real estates. This development has already started during the implementation of the THIP. Speculation for land and housing led to price increases that made land less affordable to lower-income people and also resulted in rent increases (interview journalist B, C, UN-Habitat expert A; also see Mengo 2011; Standard 2012). As a result, one outcome of the THIP is the displacement of residents that can no longer afford to live in formal estates along the Thika corridor:

> "But chances are very high that, you know, that people will be evicted, because if I own a piece of land here. And my next neighbor has put up a very decent house, ah, development there and renting it out and is getting some good money, because he put up a decent development. Because I'm in business, I will also be tempted to do the same. So displacement will continue." (interview journalist C)

On the factual side, interviewees were describing the affordability of the places along the Thika Highway differently, ranging from a characterization of an area where you can still find affordable housing (interview local manufacturer, local students) to one of upper-middle income housing even before the THIP (national government bureaucrat A). Anyhow, for affected people the affordability aspect is based on a different logic:

> "(...) I mean if your income level has not considerably changed since four years, or even three years ago, and rental rates are going up where you've been living – what do you do? You look for a cheaper place. And you shift." (interview journalist C)

This means that the relative change in affordability is the crux here, not the absolute numbers. This shift can be interpreted as the functioning of Kenya's textbook capitalist market, but at the same time it is also a displacement of people on economic grounds – and this is what is happening in the THIP area. The question arises of where displaced people go. Interviewees with a strong belief in the functioning of the market system explained that these people go to places they can afford. While this is factually true, the displacement also concerns accessibility features. Several interviewees were expecting people to move (1) farther away from the highway into the hinterland, (2) farther up north with greater distance to the Nairobi Metropolitan Region, or (3) into existing informal settlements and slums at the urban fringe of Nairobi. The first option poses the problem of non-existing or sub-standard infrastructure provision and hindered access to urban services that can be found at the (new) growth centers along the highway. The second option balances out the benefits of decreased travel time and corresponding cost savings and runs against the necessity to be close to the Nairobi Metropolitan Region to access urban jobs.[50] The third option lives up to this latter criterion; however, people would experience a strong deterioration in their living conditions and also put even more pressure on the city of Nairobi and its dealing with informal settlements and illegal uses of infrastructure and basic services. The displacement is also relevant with regard to socio-economic and communal features, as one interviewee explained for lower-income people who had to move:

> "(...) they had their lives around Thika Road. They do not want to go to any other area, because they do not know where they will live. They do not know the kind of neighbors there will be, they live with. And if they will get their resources that [they] were used [to] in (...) the areas where they lived in." (interview journalist B)

The discussion about THIP's outcomes on affordable housing illustrates that the project's underlying logic of what government has to do (upgrade the road, provide necessary trunk infrastructure) and what the market can do (provide land and housing, as well as smaller-scale infrastructure) proves detrimental to lower-income people. One national bureaucrat's comment in this regard underscores once more the Kenyan Government's perception of its responsibilities in relation to the capitalist market economy:

> "So, it's very easy for people to say it's a human rights issue [to provide housing for people], but for me, I do not think the Government can do anything to stabilize the forces at play along that highway. Government has better things to do, much more urgent

---

[50] This would be further aggravated if toll stations would be introduced on Thika Highway. For further information see footnote 45.

things to do, such as first deliver education, security, and not to take care of cautioning people against rising rents in what I consider up-market neighborhoods. Those are up-market neighborhoods, they're not low-income neighborhoods, others are. (...) That's not the Government's business to do that. Why they don't take care for themselves? (...) how it's the business of the Government to compensate those people who have been hit otherwise by market forces?!" (interview national government bureaucrat A)

As was already indicated above, the displacement of lower-income people has been forced by increased real estate investment and development in upper-market housing (also see Wainaina 2012). Therefore, the changes on the other side of the THIP's outcomes on land and housing need to be presented as well.

Photo 5.10: Gated community development between Ruiru and Nairobi (Source: Author)

Upper-Market Housing

"The classy trend of healthy living within a gated community, which premiered in the American suburbs and was made popular by the hit TV series Desperate Housewives, has finally found acceptance within Kenya's middle-upper and upper classes." (Ng'Etich 2012)

"Those seeking such a life want to be around neighbours who mirror their aspirations in terms of class and lifestyle. No wonder, then, that Nairobi, Mombasa, and Kisumu are today characterised by six-foot brick walls and iron fences circling the enclaves of luxury homes, bulky gates, and 24-hour security guards to keep outsiders away." (Njiru 2013)

These two introductory quotes from newspaper articles depict the changing upper-market housing landscape of Kenya and very well describe what can and will increasingly be found along Thika Highway (also see Olingo 2012). Annex E illustrates this lifestyle living that is expressed and advertised on the websites of the plenty of new real estate developments that are already developed, currently under construction, or planned (also see Okoth 2012). If current assessments are correct and this peri-urban area is not experiencing a housing bubble (Wairimu 2013; interview investor), it will increasingly cater for what one interviewee termed a "specific class of people" (interview civil society organization representative) – meaning the rising Kenyan middle class, and probably its upper end in particular, if the steep land and housing price increases are taken into account.

With the implementation of the THIP, the area has seen gated communities being planned and constructed (Wamwari 2010; Chege 2012; Wairimu 2012c). Some interviewees described this development as gentrification (interview investor), others as segregation (interview university researcher B, C). While these terms have a much more elaborated and complex meaning in theory, their use by the interviewees hint at a type of development in the Northern Nairobi Metropolitan Region where a highway upgrade opened up peri-urban areas only to a particular group of people with specific uses and lifestyles in mind. The 'mixed-use by default' will then result in what can be called the 'real-estate/slum phenomenon' that can be found in many metropolitan areas in Kenya:

> "I think the practice in Kenya – which is not a good practice as such – but it's normal in Kenya that you always find an affluent place next to a slum. You know, a slum or an informal settlement as an adjacent place. So, I don't see that being different with the Thika Superhighway. We'll find there very high-quality property or estates along the superhighway, but of course the informal settlements along that corridor will not go away. They will still be there. So, I think they're working hand in hand, because that is the practice in several parts, particularly in Nairobi." (interview civil society organization representative)

It can be seen as a positive outcome for upper-middle class Kenyans who can now find land and housing along the THIP corridor not only available but also attractive, because the commuting to work in Nairobi or other urban areas in the metropolitan region is now easier (also see Chege 2012). At the same time, it is a missed opportunity, since the THIP triggered the same fragmented, non-integrated, and socially imbalanced development as it has happened in other urban areas of Kenya and beyond (cf. Grant 2009: 60-64). This aspect does not only concern accessibility to land and housing but also to social services that are increasingly provided along the THIP corridor. With higher-income residents in formal estates living side-by-side with their 'servants' in informal estates (at risk of being evicted), a formerly more integrated community life will likely change and represent a style of spatial segregation that was introduced through the urban studies research perspective in chapter 2.4 and which will be applied to the case study in the next section of the empirical analysis.

Photo 5.11: Boundary of the Ruiru Golf Club with trespassers sign (Source: Author).

Social Services and Community Life

With regard to positive outcomes, the THIP led to increased investment in social services, ranging from educational and medical institutions to leisure time activities and business services, such as banking and insurance (Mwongela 2010a). A second look at these new services is however insightful: It is not public social services that are improved in the THIP area; it is private sector service providers that move into this corridor due to the influx of prospective customers with the necessary spending capacity (interview UN-Habitat expert A, journalist C; also see Standard 2011).[51] It is not the objective here to criticize these service providers or their customers. It is rather about focusing on both beneficiaries and those stakeholders that are left aside in this process:

> "(…) most of it around there started off as middle income estates. But with time you're seeing a lot sophistication. We're seeing malls coming up. We're seeing big supermarkets coming up. And that one is a point that we're going to have a lot of lifestyle things which are around there. And it's only people who can afford that high standard of living that will be around there." (interview journalist C)

Existing settlements experience incremental changes to social and business services as new and often socio-economically much more privileged residents are moving in. There are examples in the research literature on possible conflicts that can arise when traditional norms, values, and needs collide with new residents' ideas of community life (Thuo 2010: 5). It is yet to be seen if the new settlements along Thika

---

[51] One interview is very revealing here, in which the investor explained the investment decision in the Southern part of the Thika Highway transport corridor closer to the already urbanized areas of the Northern Nairobi Metropolitan Region: "(…) well bear in mind that you still need to be accessible to the majority of wealthy population in that area. You know middle class is starting to access formal retail, but it's, you know, you can't ignore the rich, and we wanted to be in a position where we're closer to Muthaiga, Runda, Garden Estate, and these areas. So we didn't want to be too far out (…)."

Highway will become fragmented dormitory/commuter housing estates or vibrant settlements with a functioning community life (Mwongela 2010b). The design of the highway was already not very supportive in terms of spatial settlement features in relation to holding existing communities together:

> "You also find in some cases, (…) like you go to Ruiru, there is one neighborhood called Murera and it cuts across the road. So it has one section there, and the other section on this side. Now, of course, what happens is that they were, these were not like established estates which were cut across. There was no such thing like estate. But they were just imagined zones. (…) the road was small before, it was easy to cross, to have activities across. Now what it has done is that they're saying: 'No, you can't just cross that way now. It's not possible to cross that way.'" (interview university researcher B)

This example is very illustrative due to the interviewee's reference to "imagined zones" of community life. This shows again that the initial assessment of possible impacts of the THIP's design on existing settlements would have required a much more in-depth analysis 'on the ground' in order to develop a context-sensitive design solution. As some interviewees concluded, the engineering-bias in the project team of the THIP did not help in that regard and the aspect of accessibility, which was already inhibited on the old Thika Road, has been sacrificed in favor of mobility concerns in the THIP, thus further impinging upon community life:

> "It's a high speed road catering mainly for mobility, not accessibility! And therefore the crossing from one side to another is reduced as much as possible. (..) It means that one part of the family is on the other side and another is on this side – the coordination becomes, ah, the integration becomes much less than it used to be." (interview university researcher C)

This aspect of road design affecting community life is impossible to capture solely with technical project approaches. Proper drawings can relate to these issues, but eventually a consultative, participatory planning process with affected communities is indispensable to get to know about communal patterns and everyday uses of different parts along the Thika Highway corridor (cf. KARA & CSUD 2012). After having been constructed, the 'superhighway' is now cutting through peri-urban areas and what are considered by some people intangible social features become very much spatial and concrete, impinging on accessibility in the daily life and work of communities along the highway – depending, for instance, if footbridges or underpasses are provided or not in a particular location.

Conclusion

This aspect of bisected communities was mentioned as the final point in analyzing outcomes of the THIP with regard to land and housing, because it illustrates the interdependency of the highway design with the adjacent land and people's everyday lives. With the new upper-market housing developments along this transport corridor and the skyrocketing prices, the outcomes of the THIP are at best ambivalent, since new opportunities have been coming up for private developers and upper-middle income people, who can find housing options in that corridor that are in practical commuting distance and still significantly cheaper than those in the city of Nairobi

(Wahome 2013). At the same time, the opening up of the peri-urban area did not come with an improved, socially equal accessibility to land and housing, infrastructure, as well as social services. This exemplifies that the concept of accessibility contains many variables, one of them concerns affordability, which is put at risk in a Kenyan setting, where decision-makers and (telling from several interviews one could claim) even the majority of the public have strong believes in the functioning of a free market. Depending on how this functionality is defined, the THIP's outcomes are just as expected, as the market responded to a public investment by the government. Or, the THIP's outcomes are seen to fall short of expectations of socio-economic growth for people in that area, since transport and economic improvements are not equally shared.

Land and housing play such an essential role in Kenya – with respect to the country's history, but also people's livelihood (cf. Kingoriah 2002) – that a disregard of THIP's impacts on them has already and will increasingly result in repercussions: Instead of developing the peri-urban area strategically and sustainably, market-driven (partly speculative) investments lead to a spatial and socio-economic fragmentation of land along the THIP corridor, where infrastructure and basic services are insufficiently provided and increasingly put under pressure. Peri-urbanization poses significant challenges to planning, managing, and catering for these vast areas by the government. Nevertheless, this could/should have not come as a surprise to decision-makers behind the THIP. Right now, the Northern Nairobi Metropolitan Region is experiencing a replication of urban development failures that are well-known and characteristic of urban areas in Kenya (cf. chapter 3.2) and that are linked to questions of politics and planning in this country even beyond THIP's immediate outcomes. These interrelationships are accordingly illustrated in figure 5.5.

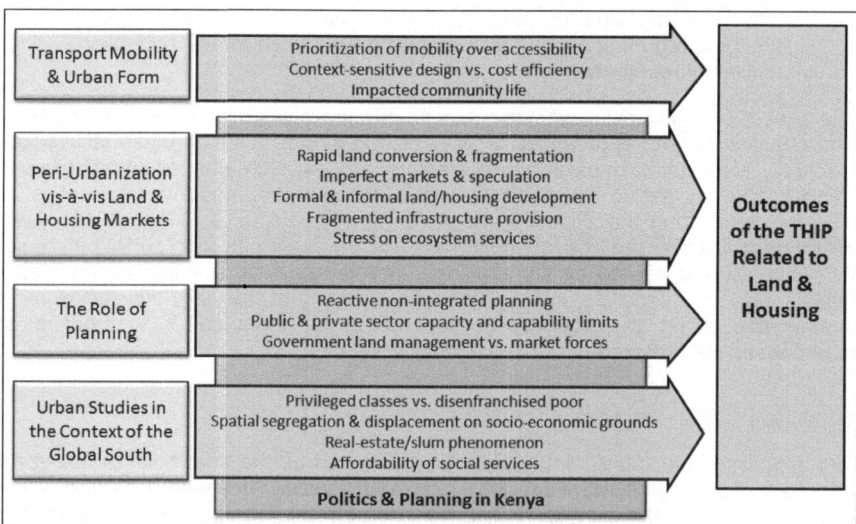

Figure 5.5: The outcomes of the THIP related to land and housing vis-à-vis relevant theoretical perspectives and their critical themes, embedded in the politics and planning context of Kenya (Source: Author).

5.5 Conclusion on Empirical Findings

> *"It's the best ever project for the country." (interview journalist A)*
>
> *"(…) I don't have anything to compare it with. (…) So the US highways are the only ones that I know, in the USA. And the others I normally see are in the movies." (interview local government bureaucrat)*
>
> *"Our people say that he who has never ventured out thinks that his mother is the best cook in the world. (…) The simple fact is that even allowing for that Nairobi-Thika 'superhighway' we are all going gaga about, our roads are abysmal. They were an insult to the science of highway engineering during the period of the cowboy contractors, and they still are an insult today when we are supposed to have kicked out all incompetent and corrupt gluttons. Nobody who have ever been to South Africa, China, Japan, North America, Europe and other parts of the world will ever suffer any illusions that we build world-class roads." (Gaitho 2012)*

Photo 5.12: Morning traffic on Thika Highway at the bottleneck entrance into Nairobi (Source: Author).

Before broader conclusions from the empirical analysis are drawn, the interviewees' conclusions on the THIP shall be compared. It is interesting to see that positive and rather negative judgments are equally represented. On the positive side, the main argument that is cited refers to the sole fact that the government actually upgraded Thika Highway, meaning that this government did not disappoint its people as many other governments had done before. Instead of a 'white elephant' a 'superhighway' was delivered:

> "Thika Highway Improvement Project can become a model. This Government has been corrupt, but one also has to acknowledge that it realized the Thika Highway project. That is its legacy. Imagine where this country would stand now if every government since independence would have built only one single functioning highway. We would not need to go to Europe for the highway system, we would feel like a developed country." (interview university researcher E)

In mixed conclusions by interviewees this achievement was often acknowledged and the necessity for the highway upgrade was not questioned. However, this group of interviewees was concerned with the question of *how* the project was delivered: They criticized the lack of context-sensitivity and were unsatisfied with the design and implementation processes. Also, there were doubts about how long some of the positive outcomes might last, given the knowledge about the long-term effect of highway expansion on induced travel (Boarnet & Chalermpong 2001: 575-576) and the challenge to introduce a proper maintenance regime to upkeep the highway's physical structure and functionality (Opukah 2012).[52] Interviewees were underscoring the lack of integration of the THIP into a larger, more holistic development plan for the (Northern) Nairobi Metropolitan Region, particularly considering the fact that 31 billion Kenyan Shillings were spent on '50 kilometers of tarmac', favoring a heavy structural measure over traffic management instruments:

> "Like in a road system, there must be what is called traffic management. (..) an expansion of a road was never a solution to congestion. (…) You expanded the road, more vehicles were coming. You expand the road, more vehicles are coming. To what extent will you continue to expand it?! (…) There's a limit. You cannot expand forever. That is why: expanding it was fine. But it should have been accompanied by proper plan of how we're going to manage the traffic, how we're going to deal with the land uses, ah, development control along the corridor – this must have come into place first!" (interview university researcher C)

There is a feeling of a "missed opportunity" (interview university researcher B) to make a real difference in the urban development along this transport corridor in contrast to other flawed projects that have been implemented or are currently under way in the country. Some interviewees concluded that the THIP represents again the failure of government to learn from past mistakes and to seriously engage with some of the broader policy issues that impact the THIP's outcomes and long-term effects, such as dealing with informal uses of land, addressing debated issues around land titles, and effectively sharing tasks between the public and the private sector in the strategic opening up of land. There were concerns that the THIP was an "over-investment" (interview national government consultant) that channeled a large amount of money into a single isolated project, thus withholding these funds from other infrastructure projects or even different policy sectors that might require similar attention.

In addition to these aspects, there seems to be a need for more reflective evaluation by decision-makers in the Kenyan Government. The THIP might be another example of Kenya's elite not committing themselves to improve the livelihood of all their people (interview university researcher A, B, national government consultant, UN-Habitat expert A), but to pursue individual interests and/or foster the 'global city' Nairobi that is characterized by exclusion and segregation, as one interviewee who works with the government reflected on:

> "(…) we, middle classers and the technocrats, and the professionals, we have (…) committed very serious errors by denying services to these informal settlements. So, as a result, you have this antagonism, which has evolved. They have failed to solve the

---

[52] On the aspect of maintenance, see footnote 45.

problem. It would have been more considerate in creating (...) a more inclusive urban development or planning. So, I think we still will have more politicized development. And it will be more contentious, because the middle class and the professionals will want to think of that world class city, that illusion of Western city and so forth...And they don't understand the context, the poverty which is underneath. And they're not doing anything about it. And that poverty will pop up and, you know, undermine all these big aspirations. (...) I think there are people who still have a dream that as a city if economy grows, (...) this poverty will disappear, and I don't believe so, because we're actually pushing more and more into that corridor." (interview university researcher B)

This broader take on issues related to the THIP was also applied in this empirical analysis by shedding light on not only the project's outcomes as intended by THIP's proponents, but also by scrutinizing outcomes in other sectors concerning different stakeholders. Therefore, it was relevant to first uncover where the project emerged from, how it was designed and eventually implemented. With the theoretical perspectives providing several critical themes to the study of a large transportation infrastructure project such as the THIP, it was possible to move beyond a narrow analysis of a road upgrade delivering smoother traffic flow and increased economic activity.

By doing so, the discussion about where the THIP emerged from already highlighted both the historic roots of the project and its interconnection with Kenya Vision 2030. It was a project that was meant not only to target the traffic issue of the Nairobi Metropolitan Region, but also to support the country's socio-economic development at a broader scale. Very common for large transportation infrastructure projects, the THIP had a strong political connotation to it, but it could be shown that the project was rationalized and born out of necessity. On the other hand, a clear interrelation between critical urban studies literature and the research on road infrastructure economics could be identified in the analysis of the THIP: It was a seemingly technical highway upgrade that was meant to pave the way for Nairobi and the country in general to follow its development path to become a middle-income country as it is imagined in Kenya Vision 2030. In direct relation to that, Nairobi Metro 2030 and the upgraded Thika Highway are in line with entrepreneurial approaches to urban development, where improved road infrastructure and enhanced mobility are key criteria for true 'world-cityness' (cf. Linehan 2007: 25, 36; Lemanski 2007: 449-450; Robinson 2006: 112-113, 139). Based on such a narrow viewpoint, only certain interests and objectives formed the foundation of the THIP.

Applied to the discussion of the THIP's design, it is illustrative to consider again the selected group of decision-makers and experts (mostly road engineers) that developed the plans for the THIP. It has been shown that a narrow set of objectives was followed and that from the beginning the project did place more emphasis on mobility than on accessibility, and favored private car use over transit options. Broader macro-economic considerations were placed above smaller-scale needs of a strategic and sustainable development of existing/emerging settlements in the peri-urban Northern Nairobi Metropolitan Region. This narrow perspective is also grounded in the process of designing and implementing the THIP that had its flaws with regard to access to information, meaningful citizen engagement, and openness to public debate. The various interests of different stakeholder groups were not considered and a strong emphasis was laid on cost-effective engineering solutions that did not take into account seemingly intangible features of existing communities in the project area, such as livelihood, community life, or affordability. Furthermore, a vertically as well as horizontally fragmented government system inhibited information sharing and cooperation to better inform the THIP. As an additional challenge, questions of metropolitan governance have not been sufficiently dealt with in the conceptionalization of this project. Insights from various research perspectives, such as critical urban studies (Lemanski 2007; Robinson 2006; Pieterse 2010b), can be informative here. They see a participatory process, which looks into the possible impacts of large public investments, as highly relevant in designing context-sensitive projects with broadly shared benefits.

It is the aspect of mitigating repercussions that particularly played out in the implementation process. Without doubt, an infrastructure project of this scale and scope does always result in some negative impacts during the construction phase. Nevertheless, the carelessness of decision-makers and the sub-standard management of the implementing companies resulted in predictable harm. There is a specific conflict reflected in the urban studies literature: While the government had the right to remove informal users from the construction area, it did not care about the affected people – instead of addressing the phenomenon, its victims were fought (which are considered as 'not innocent' or even 'guilty' by several interviewees). Informal land users were not seen as having the right to a responsible and responsive government. Thus, not enough thoughts were given to where these people would go and how they would sustain their livelihoods. This topic can only be understood in consideration of the political-historical background to the case study with regard to the Kenyan gatekeeper state with its dysfunctional planning system that can be used as a powerful instrument in a context of political corruption, high inequality, and the marginalization of a disenfranchised majority of urban dwellers (Olima 2001: 8; Kingoriah 2002: 216; UN-Habitat 2012a: 2). This highlights the relevance of unsolved policy issues, such as land, towards the implementation and outcomes of the THIP, which was not implemented in an empty space. It is the embeddedness of this large transportation infrastructure project that did receive limited attention and therefore triggered repercussions in a path-dependent way.

The empirical analysis on the THIP's outcomes illustrated best how the five applied theoretical perspectives were interlinked with each other – underscoring the complexity of what might just appear as the expansion of a road. Benefits and repercussions were equally scrutinized and the outcomes of the THIP are therefore ambivalent; specifically because clear-cut assessments are not always possible.[53] The traffic problem has been addressed, but it is questionable how long the corresponding benefits will last. There are clear job and business opportunities. Most formal and informal businesses have been benefiting from the highway upgrade. Nevertheless, a few of them became losers due to the new design of the highway that resulted in spatial features where some parts of the corridor were closed off, while new growth nodes have developed. At this point, transport mobility and urban form literature informs the understanding of how these nodes are likely to become thriving centers of different uses. On the other hand, it has been emphasized that these positive impacts depend on a system of roads and infrastructure in general that support this upgraded transport corridor. Objectives concerning inter-regional trade

---

[53] See Annex N for an overview of outcomes organized along the discussed issues in the empirical analysis and categorized as positive, ambivalent, or negative outcomes. This categorization shall be treated with caution and read in context of the argumentation in this chapter.

might be achieved, although it remains a challenge to imagine what role a 50-kilometer highway can play in relation to trans-national road systems and corresponding trade interconnectivity. This is again a question of how well the THIP has been or will be linked up with other projects and policy areas.

Turning again to the topic of agriculture, one can see how a study of outcomes in this regard brought together planning, road infrastructure economics, and peri-urbanization literature. Given agriculture's prime importance for the Northern Nairobi Metropolitan Region and beyond, this aspect might have received too little attention in the public debate and initial project evaluations. Matters of subsistence and commercial farming, food security, ecosystem balances, and preservation of arable land are fundamental to the functioning of the Nairobi Metropolitan Region and to the livelihood of its people. However, the risks emerging from rapid land conversions and corresponding loss of valuable arable land have been superseded by an equally justified focus on land and housing shortage for residential and commercial uses in the metropolitan region (Iraki 2011).[54]

The issue of land-use changes was therefore a key perspective on the THIP's outcomes and it became clear that the pace of changes in the peri-urban areas poses many challenges to the government – with regard to controlling or at least guiding development in these areas and providing infrastructure and basic services. With the inherited, still strongly centralized British planning system and a lack of cooperation between public and private actors in the strategic opening up of land, the area between Nairobi and Thika is experiencing a fragmentation that will have negative effects on other outcomes. It is not only the planning and urban studies literature that provide an understanding for this fragmentation with respect to planning and governing challenges on the regional level (UN-Habitat 2010b: 168-169; GSAPP 2006: 52-54; Neuman & Hull 2009). Research on ecosystem services also hints at the risk of peri-urban areas losing their balance and moving beyond capacity limits (UNEP 2013). Corresponding environmental impacts have not been discussed in the empirical analysis, but they can also be considered to relate to rather negative outcomes of the THIP concerning development in the peri-urban corridor area (interview journalist B, national government bureaucrat A, local government bureaucrat, civil society organization representative).

One aspect that was presented in the theoretical chapter was the role of infrastructure and its holistic planning that can be game changers in how areas grow (cf. Neumann 2009). With planning not really informing and guiding private investments, the Thika Highway corridor exemplifies an increasing spatial segregation, which is characteristic of globalizing cities (cf. Grant 2009; Lemanski 2007; Mabin 2001; UN-Habitat 2010). While land sellers, investors, developers, and the upper-middle class benefit from the mobility and access improvements of the area along the corridor, lower-income people have been displaced. The area runs the risk of establishing again a scattered, spatially segregated real-estate/slum development pattern, while former community (support) structures are lost and newly private social services are only affordable to a certain socio-economic group of people.

What was discussed in the theoretical chapter with regard to peri-urbanization vis-à-vis land and housing markets is also very well characterized by the relation of state

---

[54] See Annex O for an illustration of spatial dynamics of agriculture in peri-urban areas based on the EU-funded research project PLUREL (Piorr et al. 2011).

and market in Kenya. In addition to the government's apparent neglect to understand its responsibility for all citizens (including poor, marginalized groups), several interviewees also perceived the government as not having the necessary instruments to counter-balance the forces of a free market economy triggering rapid peri-urban land conversions in the Northern Nairobi Metropolitan Region (interview national government bureaucrat A, B, civil society organization representative, university researcher E; opposing view: local government consultant, professional body representative, national government consultant). As was already shown, this capitalist market logic does not sufficiently support socially equal accessibility to land, housing, services, etc., and it also renders the realization of THIP's intended objectives difficult, when the necessary textbook conditions of a perfect market are not given in a real-life setting that is embedded in a certain political-historical context. As a result, a seemingly technical large infrastructure project – which is thought to support economic growth, which in turn will trigger social development – actually aggravates existing problems, particularly with respect to spatial segregation on socio-economic grounds.

Therefore, one can conclude in consideration of the political-historical background of Kenya that the implied development causality linked to the THIP is not inclusive: The THIP as part of Kenya Vision 2030 is not for all Kenyans and it does not benefit all Kenyans. Therefore, it resembles flaws in the management of urban development that is typical to the Kenyan context of politics and planning. Infrastructure and planning become (re-) politicized in a situation of opening large infrastructure projects to public debate. However, this requires the political will, which is blocked by a deeply entrenched corruption and/or personal interests in the top circles of government (interview UN-Habitat expert A, university researcher A, journalist C, investor; also see KIPPRA 2005: vii; UN-Habitat 2010: 8). Nevertheless, neither the positive nor the negative conclusions on the THIP are irreversible. It was already emphasized that such a large infrastructure project is best understood as a process – and this process has not yet ended. Several outcomes have already materialized, some are hard to prevent, and others can still be influenced. Therefore, the following chapter will provide policy and research recommendations for dealing with the THIP and large transportation infrastructure projects in general, following the conclusion by one government official:

> "[The THIP] may be a lesson to learn from. For other highways in this country. Because it has its problems and mistakes." (interview national government bureaucrat A)

6. Recommendations

> *"If we cannot imagine, then we cannot manage." (Neuman & Hull 2009: 782)*

Photo 6.1: Feeder road and scattered development in the hinterland between Thika and Juja (Source: Author).

6.1 Policy Recommendations

A review of policy recommendations given in research to the topics discussed in this thesis illustrates two points: These recommendations are – even though well-minded and generally correct – often too broad to be actually ready-made for implementation in the studied cases (catchphrases: 'more democracy', 'planning for sustainability', 'equal access'). Secondly, these recommendations seem to be repeated over and over again. It is not that scholars are lacking creativity or innovation in coming up with new instruments or tools to approach urbanization challenges. Rather, it is that these policy recommendations are not rigid enough or at all implemented in practice. Therefore, it appears that the following recommendations might partly be obvious for the THIP and large infrastructure projects in general. But as was argued before, there are many variables that influence how such infrastructure projects are designed and implemented. A rather technical perspective on policy recommendations therefore needs to be applied to a specific case study and its context. Thus, the focus of this chapter lies on connecting recommendations found in research literature to the conclusions on empirical findings of this thesis – this means that the policy recommendations will be embedded in both their research and practical context.

That said, it is the aim of this chapter to propose changes to politics and planning in Kenya that are deemed feasible. The policy recommendations discussed below are again displayed in a list in Annex P, where they are evaluated towards their political, organizational, and financial feasibility. Even though broader political changes, legislative initiatives, or institutional reforms might be necessary, the following recommendations primarily focus on large transportation infrastructure projects and peri-urban areas and do not target deep transformational changes to the Kenyan state (and market) beyond this thesis' topic.

When interviewees were asked about policy recommendations following the THIP, some of them expressed their opposition to well-known blueprint 'solutions' and external lesson-teaching as there is not sufficient acknowledgement of the complexities of urbanization processes in Sub-Saharan metropolitan regions (also see Simone 2010: 34). As one government official highlighted with regard to Nairobi's situation of rapid urban growth: [55]

> "And another thing I would say is perhaps the rate of urbanization is just too high in Africa for local authorities to cope. I don't think if you have had experience with the same rates of urbanization, you would have been able to supply these [infrastructure] services. People come in their thousands every morning. What are you supposed to do? Even if you put up a hyper market, thirty stories, they will fill it up in a day, and we need to start doing another one the next day." (interview national government bureaucrat A)

The author of this thesis acknowledges these everyday practical struggles of government officials and other concerned people in the public, private, and civil society sector who are dealing with these enormous challenges, which are particularly pronounced in the Nairobi Metropolitan Region and its peri-urban areas. Therefore, the following recommendations are thought to reflect on the empirical findings of this thesis and the political-historical context in Kenya and suggest options for doing better infrastructure projects and enabling a more strategic development in the Nairobi Metropolitan Region regarding politics and planning.

6.1.1 Planning in Kenya

> "The contemporary challenge is not for planners to be able to claim expertise in each thematic area a plan might need to engage with, but rather to work productively with other professionals and equally importantly, with various bodies representing different aspects of the general public, lobby groups, interest groups, and so forth. And this is a two-way process (…)." (Allmendinger & Haughton 2009: 621-622)

This conclusion by Allmendinger and Haughton was already referred to in chapter 2.4.2 concerning new planning approaches in 'fuzzy' governance spaces. It describes how openness and readiness of planners and other experts is necessary to join forces and knowledge. A culture of information sharing, coordination, and cooperation urgently needs to be introduced in the Kenyan Government, both for horizontal and vertical styles of collaboration to break up the silo thinking that inhibits the effectiveness of project design and implementation in Kenya. In addition, the above quote hints at the necessity to train public officials in order for them to build up

---

[55] This point is also exemplified in a government document in which 150,000 urban housing units are estimated to be needed each year in Kenya with 20-30,000 only being provided (GOK 2004: 3). In a UN-Habitat document Nairobi's housing need is estimated 15,000 new units per year with only 3,000 being provided (2010: 153).

necessary capacities to deal with (changing) urbanization challenges. This is particularly relevant in a situation where lower-rank personnel might be very willing to apply state-of-the-art methods and instruments, but do not know (enough) about them (Municipal Council of Thika: 17-18; Becker 2011: 11-12).[56]

Furthermore, the Kenyan planning system requires a modernization in that large-scale structural and restrictive plans – which are most often not implemented anyway (interview journalist C, UN-Habitat expert A; also see GOK 2008b: 17) – are replaced by strategic approaches to manage urban growth. Strategic plans are less bound to force development in specific locations and instead tap into individuals' potential to best perceive where to develop land (Neumann 2009: 206). These plans can then guide the land development with supportive measures (for instance, infrastructure provision incentives for private developers). Also, these plans are more feasible, since they require less time and costs to be developed and, even more importantly, to be adjusted to changing situations on the ground (ibid.: 206). Interlinked with strategic planning approaches in Kenya would come a different perspective of public authorities on the informal and its role for the livelihood of many Kenyans (Hendriks 2010b). This is a far-reaching task that also requires a further emancipation of the Kenyan planning culture from its inherited top-down British system (see broader institutional setting below).

6.1.2 Project Design and Implementation

Applying these planning recommendations specifically to the design and implementation of projects such as the THIP, several strategic changes are advisable. To start with, studies by authors such as Beukes et al. (2011) exemplify that transportation corridors can be planned in a context-sensitive style, where the design of a road is incrementally changing depending on the location it is running through. The upgraded Thika Highway has led to some repercussions, because its design is pretty much the same from start (urban) to end (peri-urban). In contrast, any part of a road can be designed to fit into the existing fabric and to encourage interaction with adjacent settlements. In this regard, accessibility concerns of road and other infrastructure projects (referring to the previously discussed variables of physical accessibility, financial affordability, socio-economic equality, etc.) needs to be prioritized much more explicitly. For better project solutions, several design options and project alternatives have to be prepared for external evaluation (through parliamentary oversight, public debate, expert advisory committees, etc.) with information being provided in a proper and easily understandable style.

Turning to the question of whose input is sought for a project's plans, the conclusion by KARA and CSUD on the social impacts of the THIP can be referred to:

> It needs a "structured stakeholder involvement and participation at the very initial stages of project planning (...) along with continued lines of communication throughout the project circle" (2012: 20).

This is actually a clearly outlined task, which should rationally be part of properly managed infrastructure projects for which large amounts of public funds are spent. More time would be required in such participatory processes, but in light of almost

---

[56] National government bureaucrat B explained in an interview that the corresponding ministry has already been undertaking a so called "human resource capacity needs assessment" to address this issue.

inherent delays in the planning and construction of large transportation infrastructure projects, this cannot be reasonably brought to the fore as a strong counter-argument. Furthermore, projects of this nature require a political rationalization that includes open public debate on project plans that are understandable and accessible..[57]

In addition to this recommendation, appraisal reports, impacts studies, and other project plans have to be more specific about certain elements: The impact of large transportation infrastructure projects not only in the project area but in the overall region needs to be clarified to check for intra-regional imbalances and growth interdependencies (Boarnet & Haughwout 2000: 15-17; Boarnet 1998: 398). Life-cycle costs and maintenance necessities for infrastructure and basic services have to be outlined in a transparent way in order to ensure that the quality and general usability of an infrastructure piece is also given in the long-term. For identified repercussions from a project's implementation or outcomes, clear mitigation measures must be formulated and applied. It needs to be specifically described in plans how a new project is informed by lessons learnt from previous infrastructure projects.[58] The application of these lessons needs to be spelled out, and a strategy on how to learn further lessons from the new project through critical project evaluation has to be outlined.

Connecting the lesson-learning with the training of public officials (cf. International Federation of surveyors 2005: 111), the Kenyan Government should insist on making capacity building not a minor 'add-on' to the project financing by international agencies or donor countries, which are currently behind the majority of major road projects in Kenya (Becker 2011: 6; Klopp 2012b: 10). Capacity development needs to be a key part in cooperating with external project partners. At the same time, these partners would need to be required to engage with the situation on the ground instead of proposing some overly technical blueprint 'solutions' that are not helpful to develop local capacities (Hendriks 2010a: 76; Turok 2010: 15-17). In addition to these recommendations, large infrastructure projects should be required to interlink with other ongoing projects or existing programs on the national and lower levels in order to rationalize the strategic sustainability of public investments in this infrastructure project for the achievement of higher policy goals. This is a point that is informed by the previously introduced transit-oriented development (TOD) paradigm of urban development for mixed-uses (UN-Habitat 2013: Ch. 5, p. 36), its similar application in the South African context of growth management strategies that are

---

[57] As one interviewee who worked in government task forces on infrastructure projects argued: "Because the political acceptability gives you the ownership. That gives ownership! And that's the basis of it. So that people say: 'It is our project'. The project then is acceptable. When it's acceptable you can then justify the sufferings to the minority, or to the individuals. And there, the actor now who can say: 'Yeah, we can now sacrifice for this in order to move ahead.' That is the political aspect of it" (interview university researcher D).

[58] A few interviewees particularly criticized the fact that lesson-learning seems to be a big weakness of the Kenyan Government: "We don't seem to learn lessons and reflect them in the subsequent project. You know, (...) that's not our strength. We keep on doing things, I think, where we're beginners (...)" (interview local government consultant). This is also related to the Kenyan Government's limited openness to external constructive criticism/feedback on its projects: "But here, we don't do that. (...) When you bring back some research output, they think that you're criticizing them. (...) like somebody asked that question: 'I've not heard any positive you've talked about. You've only talked about negative things. You mean this road was that bad!?'...I said: 'No, but we started by talking about the positives. It is only that you have also closed your ears when we were talking about the positives. You only opened them when we're now talking about the negative. Because you're convinced there is nothing negative about the road, there are only positive things about the road'" (interview university researcher C).

interconnected to infrastructure investment plans (Todes 2012a, b), and the general approach of tying budget planning to physical planning through the strategic use of infrastructure (Neumann 2009: 208).[59]

6.1.3 Land Development in Peri-Urban Areas

As the case of the THIP exemplifies, once a road is constructed, later changes to its physical aspects are possible only to a certain extent. Footbridges, for instance, have been installed afterwards, and changes to the road design (signage, directions and turns, etc.) have been implemented to improve road safety. Nevertheless, it is clear that the focus now lies on the development of the land adjacent to the THIP, because positive as well as negative changes to these peri-urban areas are still possible.

To preserve the precious arable land along Thika Highway, the government could acquire private farmland and correspondingly compensate farmers (interview local government consultant, professional body representative, national government bureaucrat B). It could also subsidize farming while supporting a commercialization and increase in farm production (interview university researcher D). Since there is a strong need for affordable housing in the Nairobi Metropolitan Region at the same time, welfare-based demands on land are very high, thus different, partly opposing interests need to be balanced. The government could, for instance, acquire private land and set it aside in a public land bank from which then land could be released for affordable prices following strategic plans for the area (interview local government consultant, professional body representative). These tools are relatively daring given the tense setting of land policies in Kenya.[60] Therefore, Hendriks' recommendations on enabling hybrid land access could be helpful (2010b: 316). He studied peri-urban land access in the Nairobi Metropolitan Region and concluded that individuals that join forces and savings have opportunities to acquire land through cooperatives, trusts, or societies (also see Letiwa 2012). These tools can be seen as enabling land access, while informal or unregulated uses on particular parcels inside cooperative-acquired land would need to be accepted (Hendriks 2008; Rakodi 2002: 27-29). There is this challenge of finding a compromise between site-specific 'solutions' to land development and infrastructure provision and larger-scale activities of urban growth management following formal settings of government regulations (ibid.; cf. UN-Habitat 2009: 64-65). Applied to the THIP transport corridor, it can be assumed that strict zoning is unlikely to function in the Kenyan context given the above-described lack of implementation. However, more guidance to the development of land along Thika Highway could be provided if the government decides to focus on the existing and emerging growth nodes in its provision of infrastructure and basic services.

---

[59] Recall the example of the Kenya Municipal Program (see The Design of the THIP on the process of designing the THIP) that has been implemented in Thika without linking it up with the THIP, although they concerned overlapping policy areas (interview national government bureaucrat A).

[60] One interviewee's description of how compulsory acquisition by the Government is perceived in Kenya is very telling: "(...) giving the Kenyan way of looking at land, nobody will release land. Our primary schools now are choked. (..) nobody wants to give you more land. Now that means if ever these schools would have to expand, compulsory acquisition. Condemnation will have to be made out of it. And the public needs to be taught and appreciate that the Government is not condemning law if it is required. (...) that's why the Constitution allows it (...). Compulsory Acquisition is not punishment! (...) So the lessons are: a lot of public information, we need to do that, so that people appreciate the need for public projects. When they appreciate the need for public projects and maybe out of their own volition they would say: 'Can I sell? As long as you're paying me well enough?'" (interview university researcher D).

In this setting – as was already argued before – strict regulation and punitive measures against private actor development of land and infrastructure are not helpful. Instead, private developers could be offered incentives for providing affordable housing and infrastructure beyond the scale and scope of government's current financial and human resources.[61] For this approach to function, proper information needs to be provided. It is this aspect of accessibility and reliability of data on land (uses) that is essential to inform private actors on how to best develop land (Dowall 2003: 18). If the government can provide this kind of information, it can better assess what public welfare uses land should be put into and what it might have to offer to individuals to correspondingly develop certain parts of an area but not others (Olima & Kreibich 2002: 6).

With regard to the relevance and changing nature of land as a (speculative) investment, one further recommendation is highly relevant to the Kenyan context and links large infrastructure projects such as the THIP to the sustainable land development in peri-urban areas: The national government should introduce a land value capture tax. Such a tax makes clear sense, since it would help to soften to some extent speculative land acquisition, holding, and selling. Furthermore, it would cross-finance infrastructure projects that either trigger land value increases in the first place or that become necessary as land is increasingly put into use (interview local government consultant, professional body representative, national government consultant; also cf. UN-Habitat 2009: 64-65).[62]

6.1.4 Broader Institutional Setting

As the recommendation to introduce a land value capture tax exemplifies, both national and lower levels of government are called upon to work on sustainable development in the Nairobi Metropolitan Region as well as all other urbanizing areas in Kenya. For this venture, government officials (both elected politicians and bureaucrats) have to develop an understanding of their responsibility for all Kenyans – no matter if they rely on formal or informal residential and/or commercial means of living. In addition, all levels of government would need to develop more pro-active forward-thinking. Partly even the THIP exemplifies how traffic congestion was meant to be addressed by a road upgrade instead of developing a cutting-edge transport system that could have supported and guided the future growth of the Nairobi Metropolitan Region with affordable transit options and improved accessibility for various socio-economic groups of urban dwellers.

Bringing government closer to its people can prove useful here and with the ongoing implementation of the new Constitution (cf. GOK 2010) the right changes have been initiated. It goes without saying that certain powers and the necessary financial resources need to be part of this devolutionary process (cf. Wekwete 1997). What becomes apparent in the case of the Nairobi Metropolitan Region is that additional governance approaches will be needed to manage the urban agglomeration, since its administrative structure is in contrast to the idea of 'authority matching territory' (interview national government bureaucrat B; also cf. GOK 2008b; UN-Habitat 2013:

---

[61] This approach was welcomed by several interviewees, among them the private market actors, although one has to be cautious, since their approach still remains based on business considerations (i.e. costs and benefits).

[62] The introduction of toll stations would also help in this regard, but it comes with socio-economic disadvantages (see footnote 45).

Ch. 9).[63] Therefore, a policy recommendation is to set up a high-level coordinating committee on metropolitan development issues for Nairobi including all concerned counties (and national ministries if necessary). This committee would be supported by lower-rank working groups for specific topics, such as land, water, infrastructure, waste, etc. Various forms for such coordination are possible, ranging from more formalized and regularized settings to informal ways of collaboration, which partly already exist. This approach becomes even more relevant, since the responsibility for policy issues listed under Schedule Four of the new Constitution (such as land and transport; GOK 2010: 176-177) are shifted from national to county governments (Fortunate 2013). The management of the peri-urban areas in the Nairobi Metropolitan Region, thus, requires the new Nairobi County and the new Kiambu County to act in concert across administrative boundaries. The target of their work should be the strategic development of a poly-nucleated metropolitan region, where Nairobi's surrounding towns are more than the (poor) periphery to the over-dominating (congested) center.

A more incremental change related to such coordinating groups would be an increased interaction of lower-level government with actors of the private sector and civil society – here, the Second Schedule of the Urban Areas and Cities Act (GOK 2011b: 38-40) specifically mentions elements from which such participatory processes can be developed. But attention has to be paid to the fact that non-governmental organizations (and similar institutions and groups) do not necessarily represent a more democratic approach and better representation of 'the people' (interview UN-Habitat expert A, university researcher A; also Ferguson & Gupta 2002: 993). If, however, critical observation, public debate, and citizen engagement are ensured, state institutions can benefit from joining forces with other actors in addressing Nairobi's urbanization challenges in a strategic way by bundling financial and human resources (Dowall 2003: 19; Van Vliet 2002: 37; for a critical view Söderbaum & Taylor 2001: 686).

6.2   Research Recommendations

The recommendations in this part refer, first, to the study of the THIP and its outcomes and, second, to meta-level conclusions on the interrelation between research disciplines that can further enrich the understanding of the studied topics in this thesis.

With regard to the first type of recommendations, it is clear that another study of the THIP and particularly its outcomes at a later point is necessary to assess how much of this empirical analysis holds true in the future. It would also provide a closer look at the long-term effects of this large transportation infrastructure project. This is related to questions surrounding the impact of the THIP in concert with other constructed, ongoing, or planned road projects in the Nairobi Metropolitan Region, as well as beyond. A study on intra-regional urban growth effects of the THIP and Nairobi's transportation system could provide further insight. On the other hand, a study of inter-regional and trans-national road projects could shed light on aspects of trade

---

[63] This is even more the case since the last elections in March 2013, after which the Ministry of Nairobi Metropolitan Development, which had been an attempt to reify the metropolitan region, was dissolved. The new Ministry of Lands, Housing, and Urban Development has a broader, nation-wide responsibility and agenda, while tasks of the former Ministry of Nairobi Metropolitan Development should now fall under the new Nairobi County Government, although limited to its administrative boundaries.

increases, socio-economic growth along transport corridors, and enhanced regional interconnectivity.

As was argued at various points in this thesis, the study of peri-urban areas is very relevant to understand the characteristics and changes to metropolitan regions both in the Global North and South. Since peri-urban areas exhibit a distinct dynamic and great heterogeneity, research on and in these areas is rather difficult and complex. Therefore, several topics with respect to the THIP in the Northern Nairobi Metropolitan Region are still open for discussion, such as: How will the highway expansion and massive influx of people into that corridor affect ecosystem services, their capacity limits, and the region's overall balance? What can be learnt about the interdependencies of ecological functions and socio-economic development in urban agglomerations?

A different aspect that can provide increased understanding of the THIP and its outcomes is a closer study of decisions by private actors on making use of land along transport corridors. While there is an implicit thought framework based on economic theories about how private actors make investment decisions in land and housing markets, it could be enriching to look deeper into these market actions by residential developers, commercial businesses, or private social service providers in the Northern Nairobi Metropolitan Region.[64] Such an understanding is necessary to further cooperation between state and market – roles, financial and human resources, as well as expectations and interests need to be clear before a transparent collaboration can become strategic and effective.

When empirical findings from the study of the THIP and its outcomes are applied to other cases, there is also room for comparing metropolitan regions in Sub-Saharan Africa and beyond to understand what conclusions of this thesis can also be found in seemingly different contexts. It comes down to the question of whether the development of peri-urban areas can differ positively from the development of urban centers in the Global South. Or to ask more provocatively: Is empirical research on peri-urban areas able to provide insights into a socially more equal, economically more balanced, ecologically more sustainable, and politically more functional development of such areas? Or will there only be a replication of urbanization failures as they are exemplified in what UN-Habitat terms the "'Global Standard Urbanization Model of the 20th Century'" (2012b: 109)?

I would see empirical research in a position to make a positive difference, under the condition that different research disciplines are open to each other and to practical engagement. This refers to the introductory discussion of Phelps and Tewdwr-Jones (2008): The planning profession has to show more awareness of geography's deep theoretical concepts of understanding and reading developments of urban, peri-urban, and rural areas in political, social, economic, and other terms. On the other hand, geographic scholars are called upon to engage with policymaking in practice and to check in how far their research can be translated into feasible policy recommendations. As some experts in the conducted interviews remarked, this interdisciplinary research concerns not only planning and geography and it also needs to be transferred into corresponding teaching curricula that open up the silo-thinking in many disciplines (also see Syagga & Kamau 2002: 339).

---

[64] Some remarks and quotes in this thesis, particularly those of the investor, the real estate agents, and the local pharmacist and manufacturer, provided some input into this topic, but as the author observed in the interviews, much more can be uncovered about underlying economic considerations of private actors along corridor areas.

	Government shall reflect in project plans on lessons learnt from previous projects and develop a strategy on how new lessons will be learnt through critical project evaluation.	++	−	+
	Government shall require project partners to integrate capacity development as a key element in infrastructure projects by engaging with on-the-ground situations.	−	o	o
	Government shall outline in project plans, how a new project interlinks with other projects and programs vis-à-vis the furthering of higher policy goals.	++	o	++
Land Development	Government shall proactively acquire private plots to preserve arable land and to develop a public land bank for strategic land development and affordable land access.	−	−	− −
	Government shall encourage land acquisitions by cooperatives, trusts, and societies.	o	o	++
	Government shall guide land development in peri-urban areas through infrastructure provision at growth nodes.	++	+	+
	Government shall offer incentives to private developers for providing infrastructure and affordable housing.	+	+	o
	Government shall develop and maintain an accessible and reliable database on land (uses).	o	− −	o
	Government shall introduce a land value capture tax.	− −	o	++
Broader Institutional Setting	Government shall change its hostility and carelessness towards informal residential and commercial uses of land and shall aim for their integration.	−	−	+
	Government shall be more pro-active and forward-thinking in addressing urbanization challenges.	+	−	o
	Government shall further implement the new Constitution with regard to decisional and fiscal devolution.	o	−	+
	Concerned counties (and national ministries if necessary) shall set up a coordinating committee for the development of the broader Nairobi Metropolitan Region supported by lower-rank working groups for specific topics.	−	+	++
	Government shall tap into alternative financial and human resources by joining forces and sharing responsibilities of the public, private, and civil society sector in participatory processes of urban development.	o	o	++

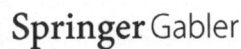

This thesis illustrated with the example of the THIP that a large transportation infrastructure project can be assessed more comprehensively when different perspectives are juxtaposed in a critical analysis. In conclusion, rather technical (economic) perspectives on such projects need to be (re-) politicized and rather political perspectives need to deal with structural/physical features of infrastructure when aiming for a comprehensive analysis of such projects and their impacts.

Since such an analysis depends on a thorough literature review, it has to be noted that some referenced literature in this thesis is categorized as 'grey literature', as it is not published in peer-reviewed academic journals. While such a differentiation makes sense, the study of the THIP and the topic of peri-urban areas underscore that 'grey literature' can provide important insights. Instead of demonizing the work of international agencies and corporations (including their publications), it is necessary to study them closely and to make use of (maybe too descriptive) analyses of corresponding topics that have not yet gained the attention of academic researchers. With the extensive appendix of this thesis and plenty of material provided online (Teipelke 2013b), I invite other researchers to further contribute to the outlined questions. I also encourage practitioners to critically engage with the findings and policy recommendations. There is more to learn on large infrastructure projects, as well as the management and governance of peri-urban areas – particularly since the livelihood of millions of urban dwellers worldwide is increasingly concerned.

7. Outlook

> *"Any urban development strategy must be built upon the aspirations, capacities, practices, and histories of a city's majority. That majority is in significant ways yet to be discovered, as it has seldom expressed or recognized itself as a coherent political entity. (…) But this notion of a majority goes beyond the statistical count or composition of electoral victories." (Simone 2010: 41)*

Photo 7.1: Thika Highway with service lanes and pedestrian walkway between Ruiru and Juja (Source: Author).

"(…) people will look at Thika Highway as a game changer (…)." (interview local government consultant)

"It's a statement! It will serve as a statement of where the people of this country want to go." (interview African Development Bank expert)

What both these quotes illustrate is the symbolic weight the THIP carries. Referring back to this thesis' introduction, it is Kenya's first 'superhighway' and as a project that was actually delivered, Kenyans can be proud of the achievement. At the same time, the THIP is embedded in Kenya Vision 2030, in which the country's elite is imagining Kenya to become a middle-income country in the next two decades. Many infrastructure projects are underway. They are transforming the life of many Kenyans, urban dwellers in particular. And hardly anyone would disagree that these projects can be improved. The THIP is relatively decent for Kenyan standards of public projects and Thika Highway is far better than most other roads in the country. Nevertheless, with the detrimental effects on livelihood and accessibility for various stakeholder groups in the Northern Nairobi Metropolitan Region, such a massive investment at least partly runs the risk of becoming another missed opportunity of making a difference in managing growth into peri-urban areas.

It is not Kenya alone that has to face the challenge of rapidly increasing urban agglomerations up to and beyond their limits. In Sub-Saharan Africa one can find many challenges related to demographics, health, employment, public safety, food security, water, in-/ formalization, etc., to be rooted in processes of peri-urbanization. Scholars such as Neuman point out: "Infrastructure systems are always planned, for better or for worse" (2009: 204). Therefore, research and practice need to actively engage with this topic, particularly, because infrastructure investments can trigger path dependencies and lock in cities in very unsustainable ways of managing growth. Kenya is still in the early stages of its demographic transformation (Hendriks 2010b: 75). Not many Nairobians even have a car, but the capital city is already highly congested (Salon & Aligula 2012: 75).

In a recent UNEP publication on city-level decoupling, the authors see urban agglomerations in the Global South as having the opportunity to learn from early mistakes of urban agglomerations in the Global North and to leapfrog to more sustainable urban growth and corresponding infrastructure solutions (2013: 86). Interviewees for this thesis give the Northern Nairobi Metropolitan Region approximately a decade to evolve into an urbanized area. As the new Constitution is incrementally implemented, the devolutionary county system has a chance to evolve to full strength. Last elections in spring 2013 were reasonably calm and brought clear results. Nairobians chose for governor a forward-thinking 'Kenya Vision 2030' candidate of the business elite instead of a polarizing person of grassroots politics. Altogether, there is a decade left to work on the Thika Highway as a project in process, as well as to make a difference in the development of the peri-urban Northern Nairobi Metropolitan Region to share benefits more broadly.

8. Bibliography

Adebayo, A. (2005): Enhancing Positive Urban-Rural Linkage Approach to Sustainable Development and Employment Generation in Southern Africa. In: UN-Habitat (Ed.): Urban-Rural Linkages Approach to Sustainable Development. Nairobi (UN-Habitat): 43-61.

ADF (2007): Appraisal Report: Nairobi-Thika Highway Improvement Project. Tunis (African Development Bank Group).

ADF (2009): Multinational Nacala Road Corridor Rehabilitation Project – Phase I: Environmental and Social Impact Assessment Summary. Tunis (African Development Bank Group).

ADF (2010a): Multinational Nacala Road Corridor Rehabilitation Project – Phase II: Environmental and Social Impact Assessment Summary. Tunis (African Development Bank Group).

ADF (2010b): Multinational Nacala Road Corridor Rehabilitation Project – Phase II: Project Appraisal Report. Tunis (African Development Bank Group).

Allmendinger, P., and G. Haughton (2009): Soft Spaces, Fuzzy Boundaries, and Metagovernance: The New Spatial Planning in the Thames Gateway. Environment and Planning A 41: 617-633.

Ancien, D. (2011): Global City Theory and the New Urban Politics Twenty Years On: The Case for a Geohistorical Materialist Approach to the (New) Urban Politics of Global Cities. Urban Studies 48 (12): 2473-2493.

Angel, S. (2011): Making Room for a Planet of Cities. Cambridge (Lincoln Institute of Land Policy).

Arnold, J. (2005): Best Practices in Corridor Management. Washington (World Bank).

Ayodo, H. (2012): Transport Upgrade Drives Property Prices. Standard: 13 September 2012. http://www.standardmedia.co.ke/?articleID=2000066004&story_title=Transport-upgrade-drives-property-prices (8 July 2013).

Ayogu, M. (2007): Infrastructure and Economic Development in Africa: A Review. Journal of African Economies 16: 75-126.

Bachmann, V. (2011): Participating and Observing: Positionality and Fieldwork Relations during Kenya's Post-Election Crisis. Area 43 (3): 362-368.

Bahati Ridge Development Ltd. (2013): Bahati Ridge Website. www.bahatiridge.co.ke (8 July 2013).

Banerjee, A., E. Duflo, and N. Qian (2012): On the Road: Access to Transportation Infrastructure and Economic Growth in China. Revised Version. National Bureau of Economic Research Working Paper No. 17897. Cambridge (NBER).

Banister, D. (2012): Assessing the Reality – Transport and Land Use Planning to Achieve Sustainability. The Journal of Transport and Land Use 5 (3): 1-14.

Barasa, L. (2012): Thika Superhighway to Steer Kenya into an Economic Hub, Says Kibaki. Daily Nation: 9 November 2012. http://www.nation.co.ke/business/news/Thika-Superhighway-to-steer-Kenya-into-an-economic-hub/-/1006/1616334/-/86eovqz/-/index.html (8 July 2013).

Baxter, J., and J. Eyles (1997): Evaluating Qualitative Research in Social Geography: Establishing 'Rigour' in Interview Analysis. Transactions of the Institute of British Geographers 22 (4): 505-525.

Bayfield, R. (2009): Destination: Argleton! Visiting an Imaginary Place. Walking Home to 50 Blog: 22 February 2009. http://walkinghometo50.wordpress.com/ 2009/02/22/destination-argleton-visiting-an-imaginary-place/ (8 July 2013).

Becker, T. (2011): Obstacles for Non-Motorized Transport in Developed Countries – A Case Study of Nairobi, Kenya. Proceedings of the European Transport Conference. Glasgow, UK: 1-15.

Bengs, C. (2005): Urban-Rural Relations in Europe. In: UN-Habitat (Ed.): Urban-Rural Linkages Approach to Sustainable Development. Nairobi (UN-Habitat): 224-242.

Beukes, E.A., M.J.W.A. Vandershuren, and M.H.P. Zuidgeest (2011): Context Sensitive Multimodal Road Planning: A Case Study in Cape Town, South Africa. Journal of Transport Geography 19: 452-460.

Boarnet, M.G. (1998): Spillovers and the Locational Effects of Public Infrastructure. Journal of Regional Science 38 (3): 381-400.

Boarnet, M.G., and A.F. Haughwout (2000): Do Highways Matter? Evidence and Policy Implications of Highways' Influence on Metropolitan Development. Paper presented at the Brookings Institution Center on Urban and Metropolitan Policy: August 2000. Washington (Brookings Institution).

Boarnet, M.G., and S. Chalermpong (2001): New Highways, House Prices, and Urban Development: A Case Study of Toll Roads in Orange County, CA. Housing Policy Debate 12 (3): 575-605.

Brenner, N., and R. Keil (Eds.) (2006): The Global Cities Reader. New York (Routledge).

Briggs, J., and D. Mwamfupe (2000): Peri-urban Development in an Era of Structural Adjustment in Africa: The City of Dar es Salaam, Tanzania. Urban Studies 37 (4): 797-809.

Buffalo Hills Leisure, Golf and Village (2013): Buffalo Hills Website. http://buffalohills.co.ke/ (8 July 2013).

Button, K. (2002): Effective Infrastructure Policies to Foster Integrated Economic Development. Paper presented at the 3rd African Development Forum: March 2002.

Buys, P., U. Deichmann, and D. Wheeler (2006): Road Network Upgrading and Overland Trade Expansion in Sub-Saharan Africa. Washington (World Bank).

Calderon Cockburn, J. (1999): Land Market in Periurban Agricultural Areas of Lima. Working Paper. Cambridge (Lincoln Institute of Land Policy).

Calderon, C. (2009): Infrastructure and Growth in Africa. Policy Research Working Paper 4914. Washington (World Bank).

Calderon, C., and L. Serven (2008): Infrastructure and Economic Development in Sub-Saharan Africa. Policy Research Working Paper 4712. Washington (World Bank).

Cervero, R. (2005): Accessible Cities and Regions: A Framework for Sustainable Transport and Urbanism in the 21st Century. Working Paper UCB-ITS-VWP-2005-3. Berkeley (University of California Berkeley Center for Future Urban Transport).

Cheboi, S. (2008a): Huge Losses as Buildings Razed. Daily Nation: 2 November 2008. http://www.nation.co.ke/News/-/1056/486574/-/tlj5c8/-/index.html (8 July 2013).

Cheboi, S. (2008b): Thika Road Demolitions Criticised. Daily Nation: 6 November 2008. http://www.nation.co.ke/News/-/1056/487968/-/tljwrg/-/index.html (8 July 2013).

Chege, N. (2012): Thika's Industrial Dreams Buried in Concrete. Standard: 6 December 2012. http://www.standardmedia.co.ke/?articleID=2000072230&story_title=Thika%E2%80%99s-industrial-dreams-buried-in-concrete (8 July 2013).

Choto, K. (2010): Utility Firms Slow Down Thika Road Project. Star: 21 October 2010. http://www.the-star.co.ke/news/article-81585/utility-firms-slow-down-thika-road-project (8 July 2013).

Consulting Engineering Services, and Runji & Partners (2011): Development of a Spatial Planning Concept for Nairobi Metropolitan Region. Study Status Executive Summary. Nairobi (Ministry of Nairobi Metropolitan Development).

Cooper, F. (2008): Africa Since 1940: The Past of the Present. New York (Cambridge University).

Crang, M.A., and I. Cook (2007): Doing Ethnographies. London (Sage).

Day, A. (2013): Kenya's Road to Development. Consultancy Africa Intelligence: 16 January 2013. http://www.consultancyafrica.com/index.php?option=com_content&view=article&id=1184:kenyas-road-to-development-&catid=58:asia-dimension-discussion-papers&Itemid=264 (8 July 2013).

Douglas, I. (2006): Peri-Urban Ecosystems and Societies: Transitional Zones and Contrasting Values. In: McGregor, D., D. Simon, and D. Thompson (Eds.): The Peri-Urban Interface: Approaches to Sustainable Natural and Human Resource Use. London/Sterling (Earthscan): 18-29.

Dowall, D.E. (2003): Land Into Cities: Urban Land Management Issues and Opportunities in Developing Countries. Paper presented at the Lincoln Institute of Land Policy Course "Comparative Policy Perspectives on Urban Land Market Reform in Eastern Europe, Southern Africa and Latin America": 7-9 July 1998. Cambridge (Lincoln Institute of Land Policy).

Durand-Lasserve, A. (2003): Land Tenure, Property System Reforms and Emerging Urban Land Markets in Sub-Saharan Africa. Paper presented at the Lincoln Institute of Land Policy Course "Comparative Policy Perspectives on Urban Land Market Reform in Eastern Europe, Southern Africa and Latin America": 7-9 July 1998. Cambridge (Lincoln Institute of Land Policy).

Earth Institute (2012): Thika Highway Improvement Project: What Lessons Learnt?. Public Forum of the Panel Discussion at University of Nairobi: 20 November 2012. http://youtu.be/Y-06qlpbbfw (8 July 2013).

El-Shakhs, S. (1997): Towards Appropriate Urban Development Policy in Emerging Mega-Cities in Africa. In: Rakodi, C. (Ed.): The Urban Challenge in Arica: Growth and Management of Its Large Cities. Online version (without pages). Tokyo (United Nations University). http://archive.unu.edu/unupress/unupbooks/uu26ue/uu26ue00.htm (8 July 2013): Ch. 14.

Englebert, P. (1997): The Contemporary African State: Neither African Nor State. Third World Quarterly 18 (4): 767-775.

Etherington, D., and M. Jones (2009): City-Regions: New Geographies of Uneven Development and Inequality. Regional Studies 43 (2): 247-265.

Ferguson, J., and A. Gupta (2002): Spatializing States: Toward an Ethnography of Neoliberal Governmentality. American Ethnologist 29 (4): 981-1002.

Flick, U. (2012): Triangulation in der qualitativen Forschung. In: Flick, U., E. von Kardoff, and I. Steinke (Eds.): Qualitative Forschung: Ein Handbuch. 9th ed. Hamburg (Rowohlt): 309-318.

Fortunate, E. (2013): Governors Handed More Powers after State House Talks. Daily Nation: 19 June 2013. http://www.nation.co.ke/News/politics/Governors-handed-more-powers/-/1064/1888784/-/n74qkf/-/index.html (8 July 2013).

Gaitho, M. (2012): Let's Not Gloat about Building Great Roads until We Venture out of Kenya. Daily Nation: 25 June 2012. http://www.nation.co.ke/oped/ Opinion/-/440808/1435452/-/lqcdb7z/-/index.html (8 July 2013).

Geita, M.(2012): Doomsayers Should Stop Misleading Kenyans about Thika Superhighway. Standard: 24 June 2012. http://www.standardmedia.co.ke/? articleID=2000060466&story_title=Doomsayers-should-stop-misleading-Kenyans-about-Thika-superhighway (8 July 2013).

Gisesa, N. (2012): Nightmare of Navigating Super Highway. Daily Nation: 21 January 2012. http://www.nation.co.ke/News/Nightmare-of-navigating-super-highway -/-/1056/1311710/-/jopyai/-/index.html (8 July 2013).

Government of Kenya (2004): National Housing Policy for Kenya. Sessional Paper No. 3. Nairobi (Ministry of Housing).

Government of Kenya (2007): Kenya Vision 2030 – Popular Version. Nairobi (Government of Kenya).

Government of Kenya (2008a): First Medium Term Plan, 2008-2012. Nairobi (Office of the Prime Minister/Ministry of State for Planning, National Development and Vision 2030).

Government of Kenya (2008b): Nairobi Metro 2030 – A World Class African Metropolis: Building a Safe, Secure and Prosperous Metropolitan. Nairobi (Ministry of Nairobi Metropolitan Development).

Government of Kenya (2010): The Constitution of Kenya. Nairobi (National Council for Law Reporting).

Government of Kenya (2011a): Kenya Vision 2030 Website. http://www.vision2030.go.ke/ (8 July 2013).

Government of Kenya (2011b): The Urban Areas and Cities Act. Nairobi (National Council for Law Reporting).

Grant, R. (2009): Globalizing City: The Urban and Economic Transformation of Accra, Ghana. Syracuse University (Syracuse).

GSAPP (2006): Nairobi: Metropolitan Expansion in a Peri-urban Area. Studio Report. New York (Graduate School of Architecture, Planning and Preservation/ Columbia University).

Guangyuan, L. (2012): Thika Superhighway the Ultimate Emblem of Sino-Kenyan Friendship. Standard: 10 November 2012. http://www.standardmedia.co.ke/ ?articleID=2000070359&story_title=Thika-Superhighway-the-ultimate-emblem-of-Sino-Kenyan-friendship (8 July 2013).

Harding, A. (2007): Taking City Regions Seriously? Response to Debate on 'City-Regions: New Geographies of Governance, Democracy and Social Reproduction'. International Journal of Urban and Regional Research 31 (2): 443-458.

Harper, D. (2012): Fotografien als sozialwissenschaftliche Daten. Trans. Alexandre Métraux. In: Flick, U., E. von Kardoff, and I. Steinke (Eds.): Qualitative Forschung: Ein Handbuch. 9th ed. Hamburg (Rowohlt): 402-416.

Harrison, J. (2007): From Competitive Regions to Competitive City-Regions: A New Orthodoxy, but some Old Mistakes. Journal of Economic Geography 7: 311-332.

Harrison, J. (2010): Networks of Connectivity, Territorial Fragmentation, Uneven Development: The New Politics of City-Regionalism. Political Geography 29: 17-27.

Helfferich, C. (2011): Die Qualität qualitativer Daten: Manual für die Durchführung qualitativer Interviews. 4th ed. Wiesbaden (VS Verlag).

Hendriks, B. (2008): The Social and Economic Impacts of Peri-Urban Access to Land and Secure Tenure for the Poor: The Case of Nairobi, Kenya. International Development Planning Review 30 (1): 27-66.

Hendriks, B. (2010a): City-Wide Governance Networks in Nairobi: Towards Contributions to Political Rights, Influence and Service Delivery for Poor and Middle-Class Citizens?. Habitat International 34: 59-77.

Hendriks, B. (2010b): Urban Livelihoods, Institutions and Inclusive Governance in Nairobi: 'Spaces' and their Impacts on Quality of Life, Influence and Political Rights. PhD Thesis. Amsterdam (University of Amsterdam).

Homeafrika Communities Ltd. (2013): Migaa Website. http://migaa.com/ (8 July 2013).

International Federation of Surveyors (2005): Urban-Rural Interrelationship for Sustainable Development. In: UN-Habitat (Ed.): Urban-Rural Linkages Approach to Sustainable Development. Nairobi (UN-Habitat): 101-114.

Iraki, Xn. (2011): What Are the Economic Benefits of Thika Superhighway?. Standard: 19 July 2011. http://www.standardmedia.co.ke/?articleID =2000039166&story_title=What-are-the-economic-benefits-of-Thika-superhighway (8 July 2013).

ITDP, and EMBARQ (2012): The Life and Death of Urban Highways. New York/Washington (Institute for Transportation and Development Policy/EMBARQ).

JICA (2006): The Study on Master Plan for Urban Transport in the Nairobi Metropolitan Area in the Republic of Kenya. Final Report. Executive Summary. Nairobi (Japan International Cooperation Agency).

Jonas, A.E.G. (2012): City-Regionalism: Questions of Distribution and Politics. Progress in Human Geography 36: 822-829.

Jonas, A.E.G., and K. Ward (2007): Introduction to a Debate on City-Regions: New Geographies of Governance, Democracy and Social Reproduction. International Journal of Urban and Regional Research 31 (1): 169-178.

Kabukuru, W. (2012): LAPSSET Components. LAPSSET Tracker: 11 April 2012. http://lapssettracker.blogspot.de/2012/04/lapsset-components.html (8 July 2013).

Kahumba, P. (2011): How the Proposed Greater Southern Bypass Will Affect Nairobi Park. Fonnap Blog: 23 August 2011. http://fonnap.wordpress.com/2011/08/23/how-the-proposed-greater-southern-bypass-will-affect-nairobi-park/ (8 July 2013).

K'Akumu, O.A., and W.H.A. Olima (2007): The Dynamics and Implications of Residential Segregation in Nairobi. Habitat International 31: 87-99.

Kamau, M. (2013): New Shopping Mall Expected to Create over 1,000 Jobs. Standard: 10 January 2013. http://www.standardmedia.co.ke/?articleID

=2000074679&pageNo=2&story_title=New-shopping-mall-expected-to-create-over-1,000-jobs (8 July 2013).

Kante, B. (2005): The Urban-Rural Environment Linkages: Integrating the Brown and Green Environment Agendas. In: UN-Habitat (Ed.): Urban-Rural Linkages Approach to Sustainable Development. Nairobi (UN-Habitat): 244-257.

Kanyangi, E.O. (2010): The Impact of Thika Highway Expansion on Real Estate Property Values with Special Reference to Land Values: A Case Study of Ruiru Town. Research Project, No. 604. Nairobi (Institution of Surveyors of Kenya).

KARA, and CSUD (2012): Thika Highway Improvement Project: The Social/Community Component of the Analysis of the Thika Highway Improvement Project. Nairobi (Kenya Alliance of Resident Associations/ Center for Sustainable Urban Development).

Kibaki, M. (2012): Speech by His Excellency Hon. Mwai Kibaki, C.G.H., M.P., President and Commander-in-chief of the Defence Forces of the Republic of Kenya on the Occasion of the Official Opening of Nairobi-Thika Superhighway, 9th November 2012. Nairobi (Office of the President).

Kihanya, M. (2013): Can Toll Fees Maintain Road?. Daily Nation: 28 April 2013. http://www.nation.co.ke/Features/lifestyle/Can-toll-fees-maintain-road/-/1214/1759592/-/lw8fptz/-/index.html (8 July 2013).

Kingoriah, G.K. (2002): The Implications of Political Will on Planning Policy and Land-use in Kenya. In: Kreibich, V., and W.H.A. Olima (Eds.): Urban Land Management in Africa. SPRING Centre (Dortmund): 212-218.

KIPPRA (2005): Urban and Regional Planning as an Instrument for Wealth and Employment Creation. Proceedings of the National Conference. Nairobi, Kenya. Nairobi (Kenya Institute for Public Policy Research and Analysis).

Klopp, J. (2000): Pilfering the Public: The Problem of Land Grabbing in Contemporary Kenya. Africa Today 47 (1): 7-26.

Klopp, J. (2012a): Some Serious Lessons from Thika Highway. Nairobi Planning Innovations Blog: 21 December 2012. http://nairobiplanninginnovations.com/2012/12/21/some-serious-lessons-from-thika-highway/ (8 July 2013).

Klopp, J. (2012b): Towards a Political Economy of Transportation Policy and Practice in Nairobi. Urban Forum 23: 1-21.

Knight Frank (2012): Kenya Market Update. 2nd quarter report. Nairobi (Knight Frank Kenya).

Kreibich, V., and W.H.A. Olima (2002): Urban Land Management in Africa. SPRING Centre (Dortmund).

Kumar, A., and F. Barrett (2008): Stuck in Traffic: Urban Transport in Africa. Africa Infrastructure Country Diagnostic. Draft Final Report. Washington (World Bank).

Kumba, S. (2010): Scramble for Plots Near Road By-Passes. Daily Nation: 16 August 2010. http://www.nation.co.ke/News/Scramble-for-plots-near-road-by-passes-/-/1056/977900/-/sqis4ez/-/index.html (8 July 2013).

Kwama, K. (2013): Africa Hobnobs with China to Steer Growth. Standard: 15 January 2013. http://www.standardmedia.co.ke/?articleID=2000075055&story_title=Africa-hobnobs-with-China-to-steer-growth (8 July 2013).

Lambert, A. (2011): The (mis)measurement of periurbanization. Trans. Eric Rosencrantz. Metropolitics: 11 May 2011. http://www.metropolitiques.eu/The-mis-measurement-of.html (8 July 2013).

Lamnek, S. (2010): Qualitative Sozialforschung. 5th ed. Weinheim/Basel (Beltz).

Lemanski, C. (2007): Global Cities in the South: Deepening Social and Spatial Polarisation in Cape Town. Cities 24 (6): 448-461.

Letiwa, P. (2012): Investment Clubs Emerge as Feasible Option to Home Financing Problem. Daily Nation: 16 August 2012. http://www.nation.co.ke/Features/money/-/435440/1479586/-/yeswunz/-/index.html (8 July 2013).

Linehan, D. (2007): Re-Ordering the Urban Archipelago: Kenya Vision 2030, Street Trade and the Battle for Nairobi City Centre. Aurora Geography Journal 1: 21-37.

Louis Berger Group (2005): East and Central Africa Global Competitiveness Hub: Strategies for the Transformation of the Northern Corridor into an Economic Development Corridor. Amsterdam/Morristown (Bearingpoint/Louis Berger Group).

Mabin, A. (2001): Suburbs and Segregation in the Urbanizing Cities of the South: A Challenge for Metropolitan Governance in the Early Twenty-First Century. Paper presented at the Lincoln Institute of Land Policy Course "International Seminar on Segregation in the City": 26-28 July 2001. Cambridge (Lincoln Institute of Land Policy).

Mairura, S.M. (2010): An Analytical Study of the Effects of Highway Improvement Projects on the Property Market and Values. Research Project, No. 587. Nairobi (Institution of Surveyors of Kenya).

Mandel, J.L. (2003): Negotiating Expectations in the Field: Gatekeepers, Research Fatigue and Cultural Biases. Singapore Journal of Tropical Geography 24 (2): 198-210.

Mathenge, O. (2008): Thika Road Expansion to Cost Sh24 billion. Daily Nation: 22 August 2008. http://www.nation.co.ke/News/regional/-/1070/462046/-/6jq1p7/-/index.html (8 July 2013).

Mathiu, M. (2012): Modern Highways Are not Enough; We also Need to Train Drivers Afresh. Daily Nation: 1 March 2012. http://www.nation.co.ke/blogs/Modern-highways-are-not-enough/-/446718/1357118/-/view/asBlogPost/-/9jafv4z/-/index.html (8 July 2013).

Mayring, P. (2002): Einführung in die Qualitative Sozialforschung: Eine Anleitung zu qualitativem Denken. 5th ed. Weinheim/Basel (Beltz).

Mayring, P. (2010): Qualitative Inhaltsanalyse. 11th ed. Weinheim/Basel (Beltz).

Mbiba, B., and M. Huchzermeyer (2002): Contentious Development: Peri-Urban Studies in Sub-Saharan Africa. Progress in Development Studies 2 (2): 113-131.

McGregor, D:, D. Simon, and D. Thompson (2006): The Peri-Urban Interface: Approaches to Sustainable Natural and Human Resource Use. London/Sterling (Earthscan).

McGuirk, P. (2007): The Political Construction of the City-Region: Notes from Sydney. International Journal of Urban and Regional Research 31 (1): 179-187.

McGuirk, P., and R. Dowling (2011): Governing Social Reproduction in Masterplanned Estates: Urban Politics and Everyday Life in Sydney. Urban Studies 48 (12): 2611-2628.

MegaProjects Kenya (2012): Thika Road Blog. http://thikaroad.blogspot.de (8 July 2013).

Mengo, B. (2011): Superhighway Dream Haunts City Residents as Rents Set to Shoot Up. Daily Nation: 16 December 2011. http://www.nation.co.ke/News/-/1056/1290838/-/10cbocnz/-/index.html (8 July 2013).

Merriam-Webster Dictionary (2013a): Accessibility. http://www.merriam-webster.com/dictionary/accessibility (8 July 2013).

Merriam-Webster Dictionary (2013b): Mobility. http://www.merriam-webster.com/dictionary/mobility (8 July 2013).

Muiruri, P. (2012): The Curse of Thika Road. Standard: 15 November 2012. http://www.standardmedia.co.ke/?articleID=2000070655&story_title=The-curse-of-Thika-Road (8 July 2013).

Mundia, C.N., and M. Aniya (2006): Dynamics of Landuse/Cover Changes and Degradation of Nairobi City, Kenya. Land Degradation and Development 17: 97-108.

Mundia, C.N., and Y. Murayama (2010): Modeling Spatial Processes of Urban Growth in African Cities: A Case Study of Nairobi City. Urban Geography 31 (2): 259-272.

Municipal Council of Thika (n.d.): 5 Years Strategic Plan 2010-2014. Thika (Municipal Council of Thika).

Mwangi, I.K. (2002): Challenges for Urban Land-use Planning and Managing Development in the City of Nairobi and Bordering Urban Areas. In: Kreibich, V., and W.H.A. Olima (Eds.): Urban Land Management in Africa. SPRING Centre (Dortmund): 198-211.

Mwongela, F. (2010a): Thika Road Expansion Hold Promises for Ruiru. Standard: 4 March 2010. http://www.standardmedia.co.ke/?articleID=2000004722&story_title=Thika-Road-expansion-hold-promises-for-Ruiru (8 July 2013).

Mwongela, F. (2010b): Thika Town, the New Nairobi Suburb. Standard: 20 May 2010. http://www.standardmedia.co.ke/?articleID=2000009843&story_title=Thika-town,-the-new-Nairobi-suburb (8 July 2013).

Nabutola, W. (2009): Urban Dynamics in Kenya: Towards Inclusive Cities. Paper presented at the FIG (International Federation of Surveyors) Working Week 2009 in Eilat, Israel: 3-8 May 2009.

Nathan Associates (2011): Corridor Diagnostic Study of the Northern and Central Corridors of East Africa. Action Plan – Volume 1: Main Report. Arlington (Nathan Associates).

Ndoria, C.W. (2011): Factors Influencing the Current Growth of the Real Estate Market in Thika Town. Research Project, No. 680A. Nairobi (Institution of Surveyors of Kenya).

Ndulu, B.J. (2006): Infrastructure, Regional Integration and Growth in Sub-Saharan Africa: Dealing with the Disadvantages of Geography and Sovereign Fragmentation. Journal of African Economies 15: 212-244.

Neuman, M. (2009): Spatial Planning Leadership by Infrastructure: An American View. International Planning Studies 14 (2): 201-217.

Neuman, M., and A. Hull (2009): The Futures of the City Region. Regional Studies 43 (6): 777-787.

Ng'Etich, J. (2012): Exclusive Homes. Daily Nation: 29 March 2012. http://www.nation.co.ke/Features/DN2/Exclusive-homes-/-/957860/1375318/-/vyyq17z/-/index.html (8 July 2013).

Ngirachu, J. (2010): Key Experts Left Out in City Roads Expansion. Daily Nation: 22 September 2010. http://www.nation.co.ke/News/Key-experts-left-out-in-city-roads-expansion/-/1056/1016064/-/7fbkui/-/index.html (8 July 2013).

Ngirachu, J. (2012): It's not Business as Usual on New Road. Daily Nation: 10 November 2012. http://www.nation.co.ke/News/Its-not-business-as-usual-on-new-road/-/1056/1616500/-/137xeq1/-/index.html (8 July 2013).

Njau, F.W. (2010): An Assessment of Current Trends of Residential Property Market in Thika Municipality and Its Environs. Research Project, No. 620. Nairobi (Institution of Surveyors of Kenya).

Njenga, S. (2013): Sewerage Plants for Juja, Ruiru. Star: 16 January 2013. http://www.the-star.co.ke/news/article-103026/sewerage-plants-juja-ruiru (8 July 2013).

Njiru, J. (2013): Sh117,000 a month? Someone Is Making a Kill Here. Daily Nation: 7 March 2013. http://www.nation.co.ke/Features/DN2/Sh117000-a-month-Someone-is-making-a-kill-here/-/957860/1713170/-/9w86daz/-/index.html (8 July 2013).

NTV (2012): Thika Highway: A 'Regional Masterpiece'. NTV: 9 November 2012. http://www.ntv.co.ke/news2/topheadlines/financing-the-thika-super-highway/ (8 July 2013).

Obala, L.M., and M. Kimani-Mukindia (2002): Land-use Conflicts and Urban Land Management in Kenya. In: Kreibich, V., and W.H.A. Olima (Eds.): Urban Land Management in Africa. SPRING Centre (Dortmund): 163-173.

Obudho, R. A. (1997): Nairobi: National Capital and Regional Hub. In: Rakodi, C. (Ed.): The Urban Challenge in Arica: Growth and Management of Its Large Cities. Online version (without pages). Tokyo (United Nations University). http://archive.unu.edu/unupress/unupbooks/uu26ue/uu26ue00.htm (8 July 2013): Ch. 9.

Obudho, R. A., and P. Juma (2002): The Role of Urbanisation and Sub-urbanisation Processes in Urban Land Management Practices in East Africa. In: Kreibich, V., and W.H.A. Olima (Eds.): Urban Land Management in Africa. SPRING Centre (Dortmund): 34-53.

Odipo, D. (2012): Road to Progress is Paved with Hope and Peace. Standard: 19 November 2012. http://www.standardmedia.co.ke/?articleID=2000070977&story_title=Road-to-progress-is-paved-with-hope-and-peace (8 July 2013).

Odipo, D. (2013): Can Kibaki Tenure Stand Scrutiny, Test of Time?. Standard: 7 January 2013. http://www.standardmedia.co.ke/?articleID=2000074428&story_title=Can-Kibaki-tenure-stand-scrutiny,-test-of-time (8 July 2013).

Oduor, A. (2010): Drivers Feel Pain of Rough Side of Super Highway. Standard: 2 February 2010. http://www.standardmedia.co.ke/?articleID=2000002291&story_title=Drivers-feel-pain-of-rough-side-of-super-highway (8 July 2013).

Oduor, P. (2013): Stop! The Crippling State of Kenya's Killer Roads. Daily Nation: 23 April 2013. http://www.nation.co.ke/Features/DN2/Accideaths/-/957860/1755208/-/item/2/-/c70kjt/-/index.html (8 July 2013).

Office of the Deputy Prime Minister and Ministry of Local Government (2010): The Kenya Municipal Program: Environmental and Social Management Framework (ESMF). Nairobi (Government of Kenya).

Onyango, L. (2011): Cost of Road to Go Up Over Fuel Prices. Daily Nation: 12 May 2011. http://www.nation.co.ke/News/Cost-of-road-to-go-up-over-fuel-prices-/-/1056/1161468/-/857rphz/-/index.html (8 July 2013).

Okoth, J. (2012): Growing Middle Class Fuelling Demand for Luxury Housing. Standard: 13 November 2012. http://www.standardmedia.co.ke/?articleID=2000070530&story_title=Growing-middle-class-fuelling-demand-for-luxury-housing (8 July 2013).

Olima, W.H.A. (2001): The Dynamics and Implications of Sustaining Urban Spatial Segregation in Kenya: Experiences from Nairobi Metropolis. Paper presented at the Lincoln Institute of Land Policy Course "International Seminar on Segregation in the City": 26-28 July 2001. Cambridge (Lincoln Institute of Land Policy).

Olima, W.H.A. (2002): The Role of Land-use Planning in Sustainable Urban Land Management in Kenya. In: Kreibich, V., and W.H.A. Olima (Eds.): Urban Land Management in Africa. SPRING Centre (Dortmund): 54-64.

Olima, W.H.A., and V. Kreibich (2002): Land Management for Rapid Urbanisation under Poverty: An Introduction. In: Kreibich, V., and W.H.A. Olima (Eds.): Urban Land Management in Africa. SPRING Centre (Dortmund): 3-10.

Olingo, A. (2012): Highway's Super Exploits for Juja. Standard: 2 August 2012. http://www.standardmedia.co.ke/?articleID=2000063151&story_title=Highway%E2%80%99s-super-exploits-for-Juja (8 July 2013).

Ombuor, J. (2012a): Man Calls for Justice after Being Displaced from Thika Road. Standard: 16 October 2012. http://www.standardmedia.co.ke/?articleID=2000068532&story_title=Man-calls-for-justice-after-being-displaced-from-Thika-Road (8 July 2013).

Ombuor, J. (2012b): Tears and Joy along Nairobi-Thika Superhighway. Standard: 16 February 2012. http://www.standardmedia.co.ke/?articleID=2000052198&story_title=Tears-and-joy-along-Nairobi-ThikaSuperhighway (8 July 2013).

Omondi, G. (2012): Motorists Face Toll Charges for Thika Road Use. Business Daily: 5 November 2012. http://www.businessdailyafrica.com/Corporate-News/Motorists-face-toll-charges-for-Thika-Road-use-/-/539550/1612586/-/d5f9wwz/-/index.html (8 July 2013).

Opoku Nyarko, J., and O. Adu-Gyamfi (2012): Managing Peri-Urban Land Development: Building on Pro-poor Land Management Principles. Paper presented at the FIG (International Federation of Surveyors) Working Week 2012 in Rome, Italy: 6-10 May 2012.

Opukah, S. (2012): Infrastructure Projects Are Fine but Are Kenyans Prepared to Maintain Them?. Daily Nation: 22 April 2012. http://www.nation.co.ke/oped/Opinion/-/440808/1391572/-/m6sm4iz/-/index.html (8 July 2013).

Ortiz, P. (2011a): Propositive Analysis: Strategic Spatial Land Use Model for the City of Nairobi. Document for Debate. Washington (World Bank).

Ortiz, P. (2011b): The Nairobi Commuter Train Regional Development: Land-Use – Transport Integration. NamSIP Project. Presentation at the World Bank: 13 December 2011.

Otieno, D. (2012): Thika Highway to Be Officially Opened. CitizenTV: 8 November 2012. http://www.citizennews.co.ke/news/2012/local/item/5523-thika-high way-to-be-officially-opened (8 July 2013).

Otiso, K.M. (2002): Forced Evictions in Kenyan Cities. Singapore Journal of Tropical Geography 23 (3): 252-267.

Otiso, K.M. (2005): Kenya's Secondary Cities Growth Strategy at a Crossroads: Which Way Forward? GeoJournal 62 (1/2): 117-128.

Oucho, J.O. (2005): Enhancing Positive Urban-Rural Linkages Approach to Sustainable Development and Employment Generation: Experiences in Eastern and Central Africa. In: UN-Habitat (Ed.): Urban-Rural Linkages Approach to Sustainable Development. Nairobi (UN-Habitat): 62-100.

Owuor, H. (2010): European Countries Lose Prize to China in Race for Stakes in Africa. Daily Nation: 30 October 2010. http://www.nation.co.ke/News/ European-countries-lose-prize-to-China/-/1056/1043672/-/ft4135/-/index.html (8 July 2013).

Onyango, P. (2012): Thika Road Losing Its Splendour to Vandals. Standard: 26 November 2012. http://www.standardmedia.co.ke/?articleID=2000071539& story_title=Thika-road-losing-its-splendour-to-vandals (8 July 2013).

Paasche, T.F., and J.D. Sidaway (2010): Transecting Security and Space in Maputo. Environment and Planning A 42: 1555-1576.

Pala, O. (2012): Our Nation Is Buried in Pessimism. Standard: 17 December 2012. http://www.standardmedia.co.ke/?articleID=2000073025&story_title=Our-nation-is-buried-in-pessimism (8 July 2013).

Parnell, S., and D. Simon (2010): National Urbanisation and Urban Policies: Necessary but Absent Policy Instruments in Africa. In: Pieterse, E. (Ed.): Urbanization Imperatives for Africa: Transcending Policy Inertia. Cape Town (African Centre for Cities/University of Cape Town): 46-59.

Pearce-Oroz, G. (2001): Causes and Consequences of Rapid Urban Spatial Segregation: The New Towns of Tegucigalpa. Paper presented at the Lincoln Institute of Land Policy Course "International Seminar on Segregation in the City", 26-28 July 2001. Cambridge (Lincoln Institute of Land Policy)

Phelps, N.A., and A.M. Wood (2011): The New Post-Suburban Politics?. Urban Studies 48 (12): 2591-2610.

Phelps, N. A., and M. Tewdwr-Jones (2008): If Geography Is Anything, Maybe It's Planning's Alter Ego? Reflections on Policy Relevance in Two Disciplines Concerned with Place and Space. Transactions of the Institute of British Geographers 33: 566-584.

Pieterse, E. (2010a): Cityness and African Urban Development. Working Paper No. 2010/42. Helsinki (United Nations University World Institute for Development Economics Research).

Pieterse, E. (2010b): Filling the Void: Towards an Agenda for Action on African Urbanization. In: Pieterse, Edgar (Ed.): Urbanization Imperatives for Africa: Transcending Policy Inertia. Cape Town (African Centre for Cities/University of Cape Town): 6-27.

Piorr, A., J. Ravetz, and I. Tosics (Eds.) (2011): Peri-Urbanisation in Europe: Towards European Policies to Sustain Urban-Rural Futures. Synthesis Report. Copenhagen (University of Copenhagen/Academic Books Life Sciences).

Rakodi, C. (1997a): Global Forces, Urban Change, and Urban Management in Africa. In: Rakodi, C. (Ed.): The Urban Challenge in Arica: Growth and Management of Its Large Cities. Online version (without pages). Tokyo (United Nations University). http://archive.unu.edu/unupress/unupbooks/uu26ue/uu26ue 00.htm (8 July 2013): Ch. 2.

Rakodi, C. (1997b): Introduction. In: Rakodi, C. (Ed.): The Urban Challenge in Arica: Growth and Management of Its Large Cities. Online version (without pages). Tokyo (United Nations University). http://archive.unu.edu/unupress/ unupbooks/uu26ue/uu26ue00.htm (8 July 2013): Ch. 1.

Rakodi, C. (1997c): Residential Property Markets in African Cities. In: Rakodi, C. (Ed.): The Urban Challenge in Arica: Growth and Management of Its Large Cities. Online version (without pages). Tokyo (United Nations University). http://archive.unu.edu/unupress/unupbooks/uu26ue/uu26ue00.htm (8 July 2013): Ch. 11.

Rakodi, C. (1997d) (Ed.): The Urban Challenge in Arica: Growth and Management of Its Large Cities. Online version (without pages). Tokyo (United Nations University). http://archive.unu.edu/unupress/unupbooks/uu26ue/uu26ue 00.htm (8 July 2013).

Rakodi, C. (2002): Interactions between Formal and Informal Urban Land Management: Theoretical Issues and Practical Options. In: Kreibich, V., and W.H.A. Olima (Eds.): Urban Land Management in Africa. SPRING Centre (Dortmund): 11-33.

Robinson, J. (2006): Ordinary Cities: Between Modernity and Development. Routledge (New York).

Salmon, M. (2011): Urban Planning Related to Development Corridors. Internal Draft Paper, Version 3. Nairobi (UN-Habitat).

Salon, D., and E.M. Aligula (2012): Urban Travel in Nairobi, Kenya: Analysis, Insights, and Opportunities. Journal of Transport Geography 22: 65-76.

Sanchez, T.W. (2000): Land Use and Growth Impacts from Highway Capacity Increases. Center for Urban Studies. Portland State University. 1-23.

Sanchez, T.W., and T. Moore (2000): Indirect Land Use and Growth Impacts of Highway Improvements. Interim Report SPR 310/327. Salem/Washington (Oregon Department of Transportation/Federal Highway Administration).

Sangira, S. (2012): Vandals Descend on New Thika Highway. Star: 6 November 2012. http://www.the-star.co.ke/news/article-94304/vandals-descend-new-thika-highway (8 July 2013).

Sassen, S. (2006): Cities and Communities in the Global Economy. In: Neil Brenner and Roger Keil (Eds.): The Global Cities Reader. Routledge (New York): 82-88.

Sidaway, J.D. (1992): In Other Worlds: On the Politics of Research by 'First World' Geographers in the 'Third World'. Area 24 (4): 403-308.

Sidaway, J.D. (2009): Shadows on the Path: Negotiating Geopolitics on an Urban Section of Britain's South West Coast Path. Environment and Planning D 27: 1091-1116.

Simon, D. (1992): Cities, Capital and Development: African Cities in the World Economy. Belhaven (London).

Simon, D. (1997): Urbanization, Globalization, and Economic Crisis in Africa. In: Rakodi, C. (Ed.): The Urban Challenge in Arica: Growth and Management of

Its Large Cities. Online version (without pages). Tokyo (United Nations University).   http://archive.unu.edu/unupress/unupbooks/uu26ue/uu26ue 00.htm (8 July 2013): Ch. 3.

Simon, D., D. McGregor, and D. Thompson (2006): Contemporary Perspectives on the Peri-Urban Zones of Cities in Developing Areas. In: McGregor, D., D. Simon, and D. Thompson (Eds.): The Peri-Urban Interface: Approaches to Sustainable Natural and Human Resource Use. London/Sterling (Earthscan): 3-17.

Simone, A.M. (2004): For the City Yet to Come: Urban Life in Four African Cities. Durham/London (Duke University).

Simone, A.M. (2010): Infrastructure, Real Economies, and Social Transformation: Assembling the Components for Regional Urban Development in Africa. In: Pieterse, E. (Ed.): Urbanization Imperatives for Africa: Transcending Policy Inertia. Cape Town (African Centre for Cities/University of Cape Town): 28-45.

Smart, A., and J. Smart (2003): Urbanization and the Global Perspective. Annual Review of Anthropology 32: 263-285.

Söderbaum, F., and I. Taylor (2001): Transmission Belt for Transnational Capital or Facilitator for Development? Problematising the Role of the State in the Maputo Development Corridor. Journal of Modern African Studies 39 (4): 675-695.

SOFRECO (2012): PIDA: Interconnecting, Integrating, and Transforming a Continent. The Regional Infrastructure that Africa Needs to Integrate and Grow through 2040. Programme for Infrastructure Development in Africa. PIDA Study Synthesis. Clichy (SOFRECO).

Standard (2011): Tatu City Sends Ripples in Ruiru. Standard: 16 June 2011. http://www.standardmedia.co.ke/?articleID=2000037251&story_title=Tatu-City-sends-ripples-in-Ruiru (8 July 2013).

Standard (2012): The Glory and Gory side of the Superhighway. Standard: 10 September 2012.   http://www.standardmedia.co.ke/?articleID=2000065878& story_title=The-glory-and-gory-side-of-the-superhighway- (8 July 2013).

Syagga, Paul M., and Hannah Kamau (2002): Education, Training and Research for Urban Land Management Practice. In: Kreibich, Volker, and Washington H. A. Olima (Eds.): Urban Land Management in Africa. SPRING Centre (Dortmund): 335-343.

Tacoli, C. (2002): Changing rural-urban interactions in sub-Saharan Africa and their impact on livelihoods: a summary. Working Paper Series on Rural-Urban Interactions and Livelihood Strategies, No. 7. London (International Institute for Environment and Development).

Tacoli, C. (2005): Changing Urban-Rural Linkages in West Africa: Livelihood Transformations and Policy Responses. In: UN-Habitat (Ed.): Urban-Rural Linkages Approach to Sustainable Development. Nairobi (UN-Habitat): 116-127.

Tatu City Ltd. (2013): Tatu City Website. http://www.tatucity.com/ (8 July 2013).

Taylor, P.J. (2004): World City Network: A Global Urban Analysis. Routledge (New York).

Teipelke, R. (2013a): Places Like, Let's Say, Highways?!. Places Blog: 27 March 2013. http://blog.inpolis.com/2013/03/27/places-like-lets-say-highways/ (8 July 2013).

Teipelke, R. (2013b): The Thika Highway Improvement Project. http://renardteipelke.wordpress.com/research/thika-highway-improvement-project/ (8 July 2013).

Tewdwr-Jones, M., and D. McNeill (2000): The Politics of City-Region Planning and Governance: Reconciling the National, Regional and Urban in the Competing Voices of Institutional Restructuring. European Urban and Regional Studies 7 (2): 119-134.

Thika Greens Ltd. (2013): Thika Greens Website. http://thikagreens.co.ke/ (8 July 2013).

Thomas, R. (2009): Development Corridors and Spatial Development Initiatives in Africa. Paper Contribution January 2009. Washington (World Bank).

Thuo, A.D.M. (2010): Community and Social Responses to Land Use Transformations in the Nairobi Rural-Urban Fringe, Kenya. Field Actions Science Reports, Special Issue 1: 1-10.

Todes, A. (2012a): New Directions in Spatial Planning? Linking Strategic Spatial Planning and Infrastructure Development. Journal of Planning Education and Research 32 (4): 400-414.

Todes, A. (2012b): Urban Growth and Strategic Spatial Planning in Johannesburg, South Africa. Cities 29: 158-165.

Turok, I. (2010): The Prospects for African Urban Economies. Urban Research and Practice 3 (1): 12-24.

UNEP (2013): City-Level Decoupling: Urban Resource Flows and the Governance of Infrastructure Transitions. A Report of the Working Group on Cities of the International Resource Panel. Swilling, M., B. Robinson, S. Marvin, and M. Hodson. Nairobi (UNEP).

UN-Habitat (2005): Urban-Rural Linkages Approach to Sustainable Development. Nairobi (UN-Habitat).

UN-Habitat (2008): The State of the World's Cities 2008/2009: Harmonious Cities. Nairobi (UN-Habitat).

UN-Habitat (2009): Global Report on Human Settlements 2009: Planning Sustainable Cities. London (Earthscan).

UN-Habitat (2010): The State of African Cities: Governance, Inequality and Urban Land Markets. Nairobi (UN-Habitat).

UN-Habitat (2011): Infrastructure for Economic Development and Poverty Reduction in Africa. The Global Urban Economic Dialogue Series. Nairobi (UN-Habitat).

UN-Habitat (2012a): Habitat Country Programme Document – Kenya 2012-2014. Internal Draft Paper. Nairobi (UN-Habitat).

UN-Habitat (2012b): State of the World's Cities 2012/2013: Prosperity of Cities. World Urban Forum Edition. Nairobi (UN-Habitat).

UN-Habitat (2013): Global Report on Human Settlements: Planning and Design for Sustainable Urban Mobility. London (Earthscan).

Van Vliet, W. (2002): Cities in a Globalizing World: From Engines of Growth to Agents of Change. Environment and Urbanization 14 (1): 31-40.

Wahome, M. (2011): Modern Transport Plan Set to Shake Up City Property Market. Daily Nation: 2 July 2011. http://www.nation.co.ke/business/news/Modern-transport-plan-set-to-shake-up-city-property-market-/-/1006/1193208/-/l1r5bt/-/index.html (8 July 2013).

Wahome, M. (2013): Land, Housing Prices in Thika Woo High-End Home Buyers. Daily Nation: 29 April 2013. http://www.nation.co.ke/Features/smart company/Land-housing-prices-in-Thika-woo-high-end-home-buyers--/-/1226/1761214/-/ey48h6/-/index.html (8 July 2013).

Wainaina, E. (2012): High Rents Push Tenants out of Houses. Daily Nation: 30 April 2012. http://www.nation.co.ke/Features/smartcompany/High-rents-push-tenants-out-of-houses-/-/1226/1396746/-/88r291/-/index.html (8 July 2013).

Wairimu, I. (2012a): Houses Everywhere, but What Will We Eat?. Daily Nation: 20 September 2012. http://www.nation.co.ke/Features/DN2/Houses-everywhere-but-what-will-we-eat/-/957860/1511464/-/eopbitz/-/index.html (8 July 2013).

Wairimu, I. (2012b): How Long Before the Thika Road Succumbs to This Exodus?. Daily Nation: 22 November 2012. http://www.nation.co.ke/Features/DN2/-/957860/1625558/-/ckkdh2/-/index.html (8 July 2013).

Wairimu, I. (2012c): Will Thika Highway Give Rise to a 'Slumburbia?'. Daily Nation: 12 September 2012. http://www.nation.co.ke/Features/DN2/Will-Thika-highway-give-rise-to-a-slumburbia/-/957860/1504712/-/e1b4dh/-/index.html (8 July 2013).

Wairimu, I. (2013): Real Estate Bubble? In Kenya, the Foundations Stand Strong. Daily Nation: 10 April 2013. http://www.nation.co.ke/Features/DN2/Real-estate-bubble-In-Kenya-the-foundations-stand-strong-/-/957860/1744922/-/1x9fn1/-/index.html (8 July 2013).

Wamwari, E. (2010): Where Life Starts and Ends at the Steel Gate. Daily Nation: 11 August 2010. http://www.nation.co.ke/Features/DN2/Where-life-starts-and-ends-at-the-steel-gate-/-/957860/974940/-/ikdwvk/-/index.html (8 July 2013).

Wamwere, K. (2013): Let Kenyans Be More Truthful with Kibaki's Legacy. Star: 5 January 2013. http://www.the-star.co.ke/news/article-101627/let-kenyans-be-more-truthful-kibakis-legacy (8 July 2013).

Ward, K., and A.E.G. Jonas (2004): Competitive City-Regionalism as a Politics of Space: A Critical Reinterpretation of the New Regionalism. Environment and Planning A 36: 2119-2139.

Wekwete, K.H. (1997): Urban Management: The Recent Experience. In: Rakodi, C. (Ed.): The Urban Challenge in Arica: Growth and Management of Its Large Cities. Online version (without pages). Tokyo (United Nations University). http://archive.unu.edu/unupress/unupbooks/uu26ue/uu26ue00.htm (8 July 2013): Ch. 15.

World Bank (2002): Scoping Study: Urban Mobility in Three Cities – Addis Ababa, Dar es Salaam, Nairobi. SSATP Working Paper No. 70. Washington (World Bank).

Wylie, J. (2005): A Single Day's Walking: Narrating Self and Landscape on the South West Coast Path. Transactions of the Institute of British Geographers 30: 234-247.

9. Appendix

9.1   Annex A: Driving Forces and Dynamics of Peri-Urbanization

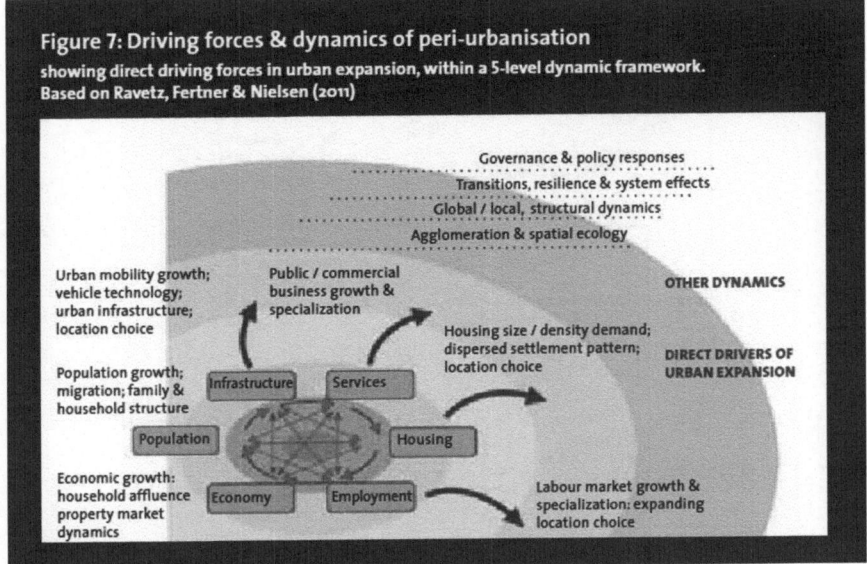

Driving forces and dynamics of peri-urbanization (Source: Piorr et al. 2011: 32).

9.2   Annex B: Spatial Land Use Changes in the Nairobi Metropolitan Region

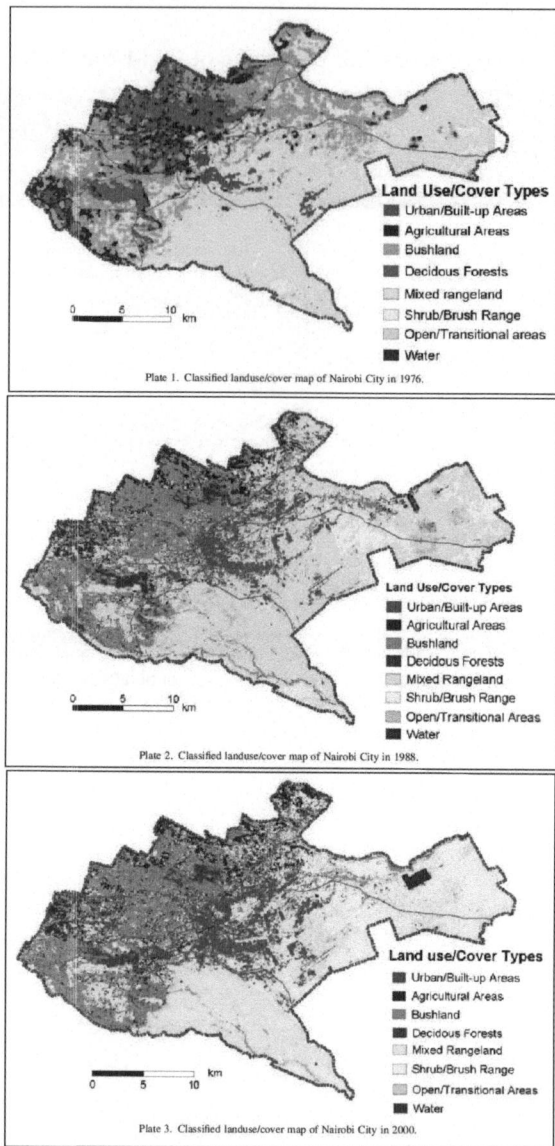

Changes of different land uses and cover types in the Nairobi Metropolitan Region from 1976 to 2000 (Source: Mundia & Aniya 2006: 105-106).

9.3   Annex C: THIP Linkage to Other Road Infrastructure Projects

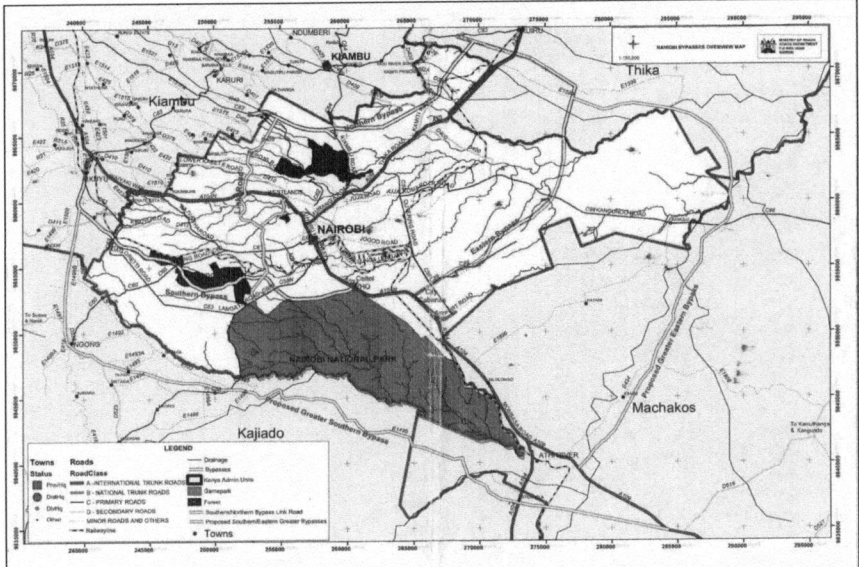

Map of proposed or already implemented bypasses for the Nairobi Metropolitan Region (Source: Kahumba 2011, based on a Government of Kenya document).

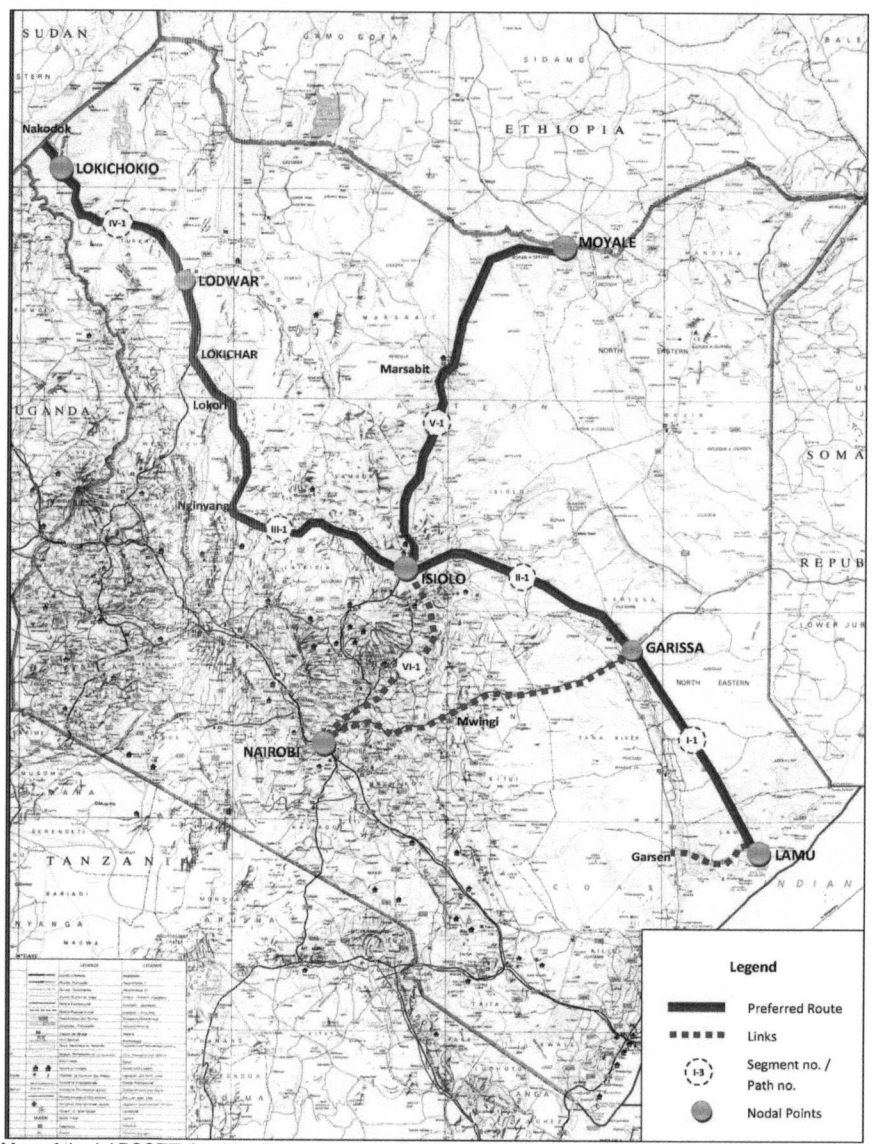

Map of the LAPSSET Corridor Route (Source: Kabukuru 2012).

9.4   Annex D: Photographic Exercise

In January and February, photos were taken during several visits to the field. After organizing the photos by naming them and deleting fuzzy images, also the contrast and other features of the photos were adjusted with the help of the software *Microsoft Picture Manager* in order to improve some of the photos' quality. Most importantly, the photos were tagged in the following categories: settlement, business, crossing, land, luxury, construction, and highway. These categories were phrased very broadly, as their main function was to organize the large stock of photos. The *settlement tag* is used for motives that show residential houses and urban structures. The *business tag* is used for photos that show businesses or business activities. The *crossing tag* stands for pictures in which material as well as human examples of crossings over the highway are exemplified. The *land tag* is put to every motive where the surrounding, non-urban area of the highway is shown. The *luxury tag* is the weakest tag as it stands for a rather fuzzy concept of high-income amenities that are somehow indicated to in the pictures. The *construction tag* is used for motives that show ongoing constructions or signs that hint at changes in the built and/or economic environment along the highway. The *highway tag* is put to every photo in which the highway is the major focus of the motive. From each category, a selection of the best photos was compiled and is available in high resolution online for free use (http://renardteipelke.wordpress.com/research/thika-highway-improvement-project/ photos/), following the regulations under the Creative Commons Attribution-NoDerivs 3.0 Unported License. The copyright of all photos lies with the author (Renard Teipelke, 2013).

The following sample of photos presents a best-of selection to provide an idea of how the study area appears.

Settlement

Medium-density housing (Nairobi to Ruiru)

Scattered peri-urban housing (Juja to Ruiru)

Scattered mixed-use settlement (Juja to Ruiru)

Permanent businesses at the town periphery (Ruiru)

Northern town entrance (Ruiru)

Expanding settlement above the highway (Witeithie)

New housing developments in the hinterland (Thika to Juja)

Side-street of an expanding settlement (Witeithie)

Growing mixed-use settlement (Juja)

Scattered densification development (Juja)

New middle-income housing developments (Nairobi to Ruiru)

Advertisement for land sales (Thika)

132

Business

Billboard for new mixed-use development (Nairobi)

Newspaper woman on Thika Highway (Ruiru to Juja)

Day laborers in front of Kenya Clay Products facility (Ruiru)

Cattle drover with cows (Thika)

Food stand and boda boda at footbridge (Juja to Thika)

Town center with diverse businesses (Thika)

Githurai market (Nairobi)

Tree nursery and hotel next to Thika Highway (Juja)

133

Small-scale manufacturer (Juja to Ruiru)

Motel next to Thika Highway (Juja)

Trading stands at an underpass of Thika Highway (Juja to Ruiru)

Maasai walking along Thika Highway (Juja to Ruiru)

Mixed businesses along Thika Highway (Ruiru)

Manufacturing shops at the town periphery (Ruiru)

Informal market area at a central matatu stop (Ruiru)

Food stands and boda boda at matatu stop (Thika to Juja)

Gas station next to Thika Highway (Nairobi)

Del Monte Pineapple Farm (beyond Thika)

Pineapple sellers at a Thika Highway access road (Thika)

Food stands and boda boda at matatu stop (Witeithie)

Crossing

Pedestrian crossing Thika Highway (Nairobi to Ruiru)

Inner-city footbridge (Nairobi)

Thika Highway passing under bridge (Nairobi to Ruiru)

Afterwards installed footbridge (Ruiru)

135

Pedestrians using underpass below Thika Highway (Juja to Thika)   Thika Highway cutting through settlement (Witeithie)

Pedestrians crossing Thika Highway instead of using footbridge (Nairobi)   Man quickly crossing Thika Highway (Nairobi)

Land

Undeveloped land next to Thika Highway (Ruiru to Juja)   Rail track at the town periphery (Ruiru)

Scattered development in the green hinterland (Juja to Ruiru)   Arable undeveloped land (Juja to Ruiru)

136

Scattered development next to cattle farming (Thika to Juja)

Ruiru River (Ruiru)

Horticulture next to Thika Highway (Thika to Juja)

Rail track through green valley (Thika to Juja)

Luxury

Billboard for Buffalo Hills real estate (Thika)

Boundary of the Ruiru Golf Club (Ruiru)

Billboards for Bahati Ridge real estate and Rainbow Ruiru Resort (Thika)

Rainbow Ruiru Resort (Ruiru)

Buffalo Hills real estate sales stall at Westgate Mall (Nairobi)

Gated community development (Ruiru to Nairobi)

Scattered low-density housing development (Thika)

Mixed-used rental apartments at Thika Highway (Juja)

Construction

Larger-scale development (Nairobi)

Gated community development (Ruiru to Nairobi)

New church under construction (Thika)

Inner-town mixed-use development (Thika)

Small shops and apartment developments (Juja to Ruiru)

Apartment development on the greenfield (Thika to Juja)

Highway

Speed pumps on Thika Highway (Juja to Thika)

Vehicles facing each other on service road (Juja to Ruiru)

Pedestrians walking along Thika Highway (Juja to Ruiru)

Kids cycling and playing on service road (Juja to Ruiru)

Stolen street signage due to vandalism (Juja to Ruiru)

Truck accident (Nairobi to Ruiru)

To-be installed toll station on Thika Highway (Juja to Ruiru)

Street signage on Thika Highway (Juja to Ruiru)

Thika Highway with matatu stop seen from footbridge (Thika to Juja)   Dual carriageway Thika Highway up north (beyond Thika)

Thika Highway with service lanes and access roads through built-up area (Nairobi)

Morning traffic at bottleneck entrance into the capital city (Nairobi)

9.5 Annex E: Real Estate Projects in the Northern Nairobi Metropolitan Region

Map of Selected Real Estate Projects in the Northern Nairobi Metropolitan Region

The following map depicts some larger real estate projects in the Northern Nairobi Metropolitan Region vis-à-vis the Thika Highway (black line) and other bypasses that are already built, under construction, or planned (orange lines). Major urban settlements along the Thika Highway are from South to North: Nairobi, Ruiru, Juja, and Thika (red rectangles).

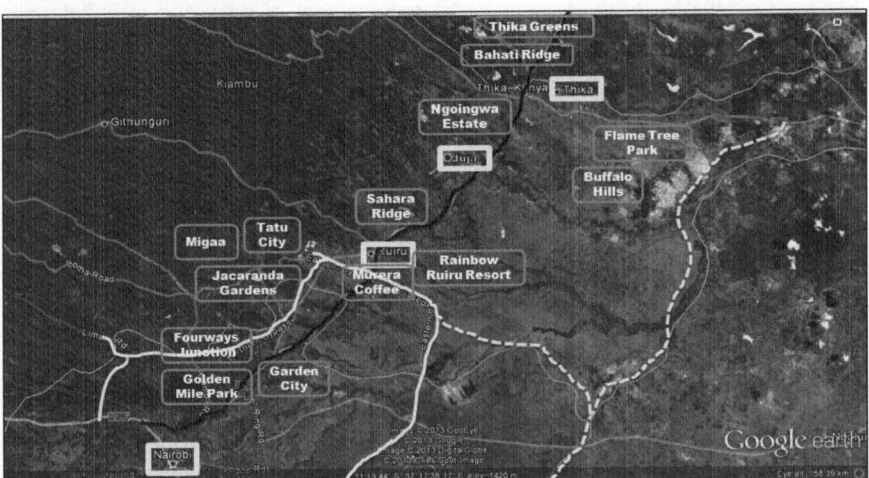

Map of Selected Real Estate Projects in the Northern Nairobi Metropolitan Region (Source: Author based on Google Earth Image 2013).

Extracts from Real Estate Websites

The following extracts from websites of selected real estate projects in the Northern Nairobi Metropolitan Region illustrate the lifestyle living that is advertised/promised to potential buyers.

Bahati Ridge (www.bahatiridge.co.ke)

"The Dream: You've dreamed of living in a country-inspired yet urban setting. You've dreamed of an elegant home in the country. You're nostalgic for the sound of cocks crowing in the morning and the warmth of wood burning fireplaces in the evening. Now you can live your dream at Bahati Ridge: your paradise is within grasp. (...) Bahati Ridge presents an exciting selection of townhouses, villas, bungalows and cottages that feature beautiful hill country views, a tranquil rural ambience and a refreshing escape from city chaos. Discerning homeowners will love this 97 acre integrated gated community and all the benefits of country living." (Bahati Ridge Development Ltd. 2013)

Buffalo Hills (http://buffalohills.co.ke/)

"Within the estate is a golf course. Well, a nine holes leisure golf course. Meaning that you can majestically tee off with your son or daughter after work, proudly walk round the golf course, discuss two or three things about school, friends, work and life in general and then walk back to your house for that sweet homely meal. You thus get time to spend with your family. After a game of golf, you can choose to catch up with your buddies at the golf club. Talk about a sweet life. On weekends, well, you do not have to worry where and with whom your kids spent their time with, they can spend time at the club house too, swim, play and grow under this soul soothing atmosphere."

"For all good things to be, there must be enforcement of rules and at buffalo hills leisure and golf village, there will be. However, this will be majorly composed by the buffalo hills golf and leisure village residents association in which, you will be a member. So you get to manage your own estate affairs. Isn't that fair?" (Buffalo Hills Leisure, Golf and Village 2013).

Migaa (http://migaa.com/)

"Away from the hustle and bustle of city life lies the magnificent Migaa"

"Migaa aims to improve the living standards of local residents by providing not only decent but also affordable housing in what will be the largest gated community in Nairobi. (…) Living at Migaa, you will find everything you and your family need to enjoy the most that life has to offer. Explore life." (Homeafrika Communities Ltd. 2013).

Tatu City (http://www.tatucity.com/)

"The vision for Tatu City is the creation of a world-class, mixed-use, mixed-income new city, located within Greater Nairobi – East Africa's new economic hub. It will provide homes and jobs for thousands of Kenyans and unparalleled economic and business opportunities."

"You can choose between the urban lifestyle and the suburban lifestyle"

"Living Your Dream: This is a city energised with opportunity, attractive, allowing freedom of movement of its citizens in complete safety. A place where children can grow up, be educated and find employment." (Tatu City Ltd. 2013).

Thika Greens (http://thikagreens.co.ke/)

"The aim of the company is to set a trail blazing path in the property development sector by delivering quality homes and world first class infrastructure."

"The security of the estate is observed, access into the estate is via gatehouses where residents have their own designated drive through the entrance, camera surveillance and police post."

"Our Lifestyle: Thika Greens is a Premier Property Investment Vehicle that creates Wholesome Lifestyles. It mission is to develop residential communities that ensures a high quality lifestyle through innovative use of resources while guaranteeing a high return on investment." (Thika Greens Ltd. 2013)

9.6   Annex F: Research for Newspaper Articles

Details on the Search Methods:

For the research of newspaper articles, Kenya's major newspaper outlets were used, including the Daily Nation, the Standard, the Star, and Business Daily. In an online search, material could be researched ranging from the most recent articles to rather old ones prior to 2009. It can be assumed that the digitalization of articles is an ongoing task for these newspaper outlets – similar to newspaper outlets in other countries. Nonetheless, it was already surprising to easily find such a large amount of articles on the topic.

The research was conducted using the corresponding search function on the outlets' websites. Search terms included, amongst others: Thika Highway, Thika Road, Thika Superhighway, Thika Highway Improvement Project, Thika, Ruiru, Juja, Transport, Infrastructure Projects, LAPSSET, LAPSET, Vision 2030.

The objective was to gain an overview of the media coverage on the topic (and related issues). Articles from other newspapers and media outlets were searched as well, although the results were very limited. Overall, the search was limited to articles in English.

The final list of newspaper articles with information on title, author, media outlet, date of publication, and link can be found on the author's website: http://renardteipelke.wordpress.com/research/thika-highway-improvement-project/photos/ (Teipelke 2013b). The list is thought to enable other researchers to further investigate the THIP and related topics.

Observations from Newspaper Articles:

There was a large variety of different writers on the THIP and related subjects. The issues covered were also relatively broad, although it can be noted that the focus laid rather on factual reporting of Government statements than on a critical assessment and reflection on the issues raised. Critical observations were equally part of the media coverage as were cheerful comments on the THIP.

With articles in 2008 mostly focusing on the demolitions that needed to be done before the actual construction of the highway started, the years 2009 up to 2011 dealt with the construction per se, the problems experienced with regard to delays and road safety, and the eviction of small traders from the road reserve or other parts of the construction site of the then-expanded highway.

The year 2012 had articles with similar topics as the years before, although with the opening in November other issues were discussed; for instance, the provision of infrastructure and basic services, the changes in design, the footbridges, and the outcomes or expected impacts of the THIP in the upcoming years. Furthermore, it was discussed how the THIP related to Kenya Vision 2030 and other transportation infrastructure projects in the country.

The most current articles from 2013 did only add few new topics, amongst them the discussion about leaving President Kibaki's legacy in light of the THIP and the country's progress as a whole. Since the second half of 2012, the issue of road maintenance has also gained momentum. Most recently, the topic of land and housing in light of increases in upper-market options has been rediscovered.

While the purely factual articles were relevant to gain information on the THIP, most revealing were the story-telling articles on individual experiences as well as the reflective articles that tried to position the THIP towards larger themes.

The number of articles was relatively equal in the years 2009 to 2011, maybe with a few more articles in 2012 with the opening approaching in the last quarter of the year. Since the THIP as a whole is currently finalized, it can be expected that there will be a less dense coverage on the highway.

9.7 Annex G: Reflections on Field Work Experience

Positionality of the Researcher:

One part of the reflection on the research process of this thesis included a reflection of the researcher's own positionality in the field and towards the research object (and interviewees, locals, etc.). Besides having been a private person experiencing Kenya, my role during the field research was double in that I was an intern with UN-Habitat and a graduate student in human geography conducting the research for my master thesis. Although based in a much more challenging research environment during the post-election violence in Kenya in 2007/2008, Bachmann's description (2011) is insightful here. He discusses his experience in a research project on Europeanization while also being affiliated with a European development agency as "occupying a double identity in the 'field' as a critical researcher and a practitioner in the development industry" (ibid.: 363), in which "(…) participation in the networks and activities of the researched can facilitate access to informants, but also necessitates carefully scrutinising the balance of the multiple roles as researcher and practitioner" (ibid.: 363) – a double role that was furthermore influenced by Bachmann being a "white, male, European researcher working in sub-Saharan Africa" (ibid.: 364). While I did not perform any active role for UN-Habitat during my research activities, my affiliation with a topic-wise relevant branch of UN-Habitat was observed and, one might even say, acknowledged by most of the people I talked to. In addition, a letter by the corresponding UN-Habitat branch asking for support to my research on the THIP opened many doors that would have otherwise been very difficult to get through (cf. Mandel 2003). Furthermore, it put me in a somewhat more authoritative power position in interviews than I would have had solely as a young white European graduate student researching a Kenyan case study (cf. Lamnek 2010: 324).

In interview situations, I sometimes had to position myself vis-à-vis UN-Habitat's and other international agencies' agendas and programs. Also in my role as a German geography student and a qualitative researcher I had to relate myself to interviewees' comments in that regard (for instance in situations where interviewees had been educated in Western/Eastern Germany themselves). The research diary was of great help here to reflect on these multiple roles. Even though the overall research process went surprisingly well – considering my limited experience in empirical research – the of my research diary below provide some insights into the setbacks or discomforts when conducting the field research, particularly during the last days of my field work in Kenya. The blog article in Annex H is another example of reflecting on my own role and perception by others during field visits.

Extracts from Research Diary:

The following four extracts from the research diary provide an impression of the author's research experience on a personal level. The diary entries are unedited and are meant to exemplify observations and reflections that were jotted down immediately after field work experiences. While the examples given below convey a rather negative picture of the experience in the field (three of the four extracts are from the last days of field work in Kenya; cf. Mandel 2003), the overall success with regard to getting things done and obtaining relevant material from the field work was surprisingly positive in the eyes of the author – given the fact that it was his first larger field research project.

12 January 2013:

"(…) Except for one older white man in a Matatu in Thika town, I was by far the only white person I have seen after leaving Nairobi. People – especially when I walked along the road from Juja to Ruiru – were very puzzled by my appearance, since they seem to have not had a clue what I was doing in this non-urban area on a Saturday. I probably did not look like someone with a specific objective when walking around. Besides, people looked like wondering when I was constantly taking pictures of the road system and adjacent features which must appear to 'locals' or bystanders as just 'concrete', 'material', 'landscape', or simple residential buildings and small businesses. Matatus as well as trucks and motorcycles passing me slowed down in several incidences asking me if I need a ride to Nairobi. One souvenir-selling Maasai who also walked along the road tried to sell me some of his stuff. Likewise, a pineapple-piece seller at the police customs control station tried to sell me his farm produce. One man in Juja told me that "we two should have a tea together", though he rather looked like one of these men you better not go with in Kenya. Nevertheless, besides this single situation I never felt unsecure or had to be afraid of anything. Especially when I came back to Nairobi, I could immediately recognize how the situation in Nairobi is much tenser and unsecure than in Thika Town or the other areas I have visited (…)."

29 January 2013:

"(…) With respect to my field visit experience, I have to underscore again that it is a very awkward situation walking along the highway and passing the settlements along the road. People seem to have not very often seen a white man in reality before – and if they have then definitely not a white man walking instead of going by matatu or, more common, car! I hear locals addressing me as "muzungu" (white person), partly friendly and welcoming, but also in a derogative sense. They are joking about me coming all the way from 'the city' and they do not have any clue why I happen to come along their settlement. There seem to be no 'reasonable reason' for this white man to make a visit to their community, even taking photos, or talking to some locals. Referring to what I have heard about field work from some experts at UN-Habitat and also researchers I talked to, it is very difficult to actually interact with locals in an effective way, since there are many hindrances: I am white, thus I am perceived as rich; not many locals do speak proper English; they do not really care about politics, do not know about their possibilities to make themselves heard, do not understand the reasons for studying the highway, do not know the concept of research, do not elaborate on their thoughts about the highway, etc. It is understandable that they are skeptical when a white man approaches them and talks gibberish (…)."

30 January 2013:

"(…) My experience today concerning the interaction with locals was overwhelmingly good and satisfying, but I am not sure how much more I can learn from locals about the highway's impact on the land along the corridor. Two people today took the time to talk to me and were very friendly and understanding when I asked them about the highway upgrade. I could easily 'catch' the pharmacist, since I had to ask her for directions and she also had no other customer coming in while we were talking. (…) The small manufacturer was sitting down on a bench right next to the small shop where we both got our soft drinks from. This was an ideal situation to engage with this man in a small talk about the highway. Since we both had to finish our bottles, before we could move on, I had sufficient time and a relaxed atmosphere to talk to him. Nevertheless, locals are not very elaborative about the highway. They have certain ideas, but they do normally not reflect on their thoughts. Often they lack interest in the politics of these issues and are not educated enough (in contrast to the two people today) to judge about certain things or to understand complex relationships impacting their daily lives. As the one expert in my interview indicated: When you engage local people in decision making, they are often surprised to hear about alternative options and the fact that they actually have a right to be heard. (...) What I also felt today is that I have increasingly a harder time dealing with the "muzungu" reactions by locals. This is because their reactions range from open-hearted interest about a white person they have probably never seen in real, to a rather hostile or derogative comment on a foreigner/stranger intruding their neighborhoods without having asked them for permission. I do not speak Swahili, thus I cannot easily interact with locals. On the other side, I am a rather sensitive person when it comes to racism, and I know from various people that Kenyans have their weaknesses in dealing with races and ethnicities in a tolerant and friendly matter. I do not feel welcomed in every situation – though I can also not judge on how I am perceived by locals. Anyhow, some of them are commenting in such a seemingly rude manner that it is hard for me to do my field work sufficiently. Maybe I have also reached a point of exhaustion, since my empirical work is so limited to a (too) tight schedule (...)."

6 February 2013:

"(...) Reaching out to locals and seeking insights from them about the highway proved to be not very effective and useful today. First of all, their English is very limited, thus an extensive conversation about something abstract like the impacts of the highway is not really possible. Furthermore, as a white person I am perceived as a stranger who really does not belong here – not in a hostile sense, but just as a funny reaction by locals towards me. When we asked our questions today, the locals in the first place did not really know how to express their opinion (in English), and in the second case, the young locals/supermarket employees in Witeithie just laughed about our questions and our appearance. The information as well as sole articulated content of these conversations does not seem to be of enough quality and/or quantity to be significantly used in my analysis and eventual master thesis (...)."

9.8 Annex H: Places Like, Let's Say, Highways?!

The following text was first published 27 March 2013 as an article on the professional blog "Places" (Teipelke 2013a).

Studying 'streets as places' through in-situ research is nothing new to social science. A classic in this regard is Doreen Massey's work on the *global sense of place* with the example of Kilburn High Road in London. Other researchers – and most naturally anthropologists – also went to the streets of settlements and studied everyday life, community, and society. But why not take the bus to go from these inner-city streets to the outer rings of urban settlements and study their larger pendants: highways?!*

It is not really the highway per se for which I spent two months at the outskirts of Nairobi, Kenya. I was studying the impact of the massive upgrade of the Thika Road – now called Thika Superhighway – on the development of the peri-urban land along this corridor. But in order to do so, I had to go to the field and 'feel' the highway. As weird this image might be, it is basically nothing else than other researchers have done when studying the impact of pedestrian-only streets on inner-city neighborhoods – it is just a rougher environment.**

In only 30 minutes, you can easily reach the highway and its adjacent towns of Ruiru, Juja, and Thika by matatu (minibus). But once you are there, you are feeling like far away from all the comforts of urban life – you better have a map in case Google Earth/Maps cannot provide precise images of the 'middle of nowhere' you are just standing in. In contrast to urban research, you need water and some food for the day, you need a lot of sunscreen, and you better know where you are going.

Even though Kenyans are extremely hospitable, residents and traders along this highway have not often seen a white person coming there without a private car. And probably even more out-of-space for them must be a white person walking (!) along the highway. If you are not a missionary – born-again and saved Christians are very welcomed in Kenya – you better have a good reason to 'visit' those peri-urban settlements. In contrast to urban in-situ research settings, these areas are rather hostile – already with regard to the physical environment, but also because residents are skeptical about 'intruders'. You can explain to them your research topic, but honestly: Who expects anyone to understand why you want to study a highway!?

Telling from my various field visits in January and February of this year, I am sure that it is nevertheless worth going into the field. Analyzing changes in the physical environment on maps and pictures or from texts is one thing. Making the reality check is another. What does it mean that a former four-lane highway is expanded to eight, ten, or in some parts even twelve lanes? How do 'vacant land', scattered development, and lack of infrastructures and services really look like? How is the interaction between people and environment – both in natural and built-up areas? And how do stark class differences and spatial segregation, which are so typical for the city of Nairobi, play out in its outskirts and satellite towns?

Those questions and many more are part of my current master thesis research, and for some of them, I will find answers. No matter how this turns out, going to the field and experiencing this rough environment and even 'feeling' the highway were necessary activities in order to know what I can actually talk and write about. A point to take home is that highways or roads through rural and peri-urban areas are also places worth to be studied. A global sense of place has already reached the 'hinterlands of globalization'…

---

* For an ethnographic perspective on this topic, check out the DFG-funded project "Roadside and travel communities. Towards an understanding of the African long-distance road".
** Walking along paths has also been a feature in *psychogeography*, particularly with reference to *transect walks*. Examples include stories/studies about urban as well as rural areas (see for instance Paasche & Sidaway 2010; Wylie 2005).

9.9   Annex I: Interview Questionnaire

Introduction

How would you describe the THIP in just a few words?

What features has this project with regard to non-motorized transport and uses?

Vision 2030

In how far is the THIP part of Kenya Vision 2030 (and Nairobi Metro 2030) and its implementation?

What role does the transportation element in Kenya Vision 2030 play?

How is the transportation element related to other elements, such as the economic element and the urban development in and around Nairobi? (grander theme?)

Actors and Interests

Which were the major actors in the THIP?

What would you say have been the main interests of these actors?

What overlapping or conflicting interests between these actors can you identify?

What other interests should have played a role in this project?

Why did those interests not play a role?

What was your role in the project? / How would you see your organization vis-à-vis that project?

Implementation Process

How would you describe the implementation process of the THIP?

What problems in this implementation process can you identify? What could have been the reasons for these problems? Could these problems be solved during the implementation process? Why?

In how far are transportation infrastructure planning and the development of the peri-urban area along the highway connected in the THIP?

Were there specific conflicts with regard to land uses?

Were there plans to set aside land for informal businesses along the road?

To what degree was the THIP based on a top-down process?

In how far lived the THIP up to expectations concerning transparency, information sharing, and community participation?

Outcomes

How would you describe the outcomes of the THIP?

What major changes will result from this project in the short/medium/long term?

Whom would you identify as the main beneficiaries of this project and why?

What other stakeholder groups could have benefited from this project? Why have they not (yet) benefitted from it?

How could they have benefitted from the project? Or how can they benefit from the project in the future? What role does 'ownership' play in this regard? What about easier credits and loans? What about land use policies, building codes, improved basic infrastructure?

Were decision makers concerned about the likely displacement of poorer residents along Thika highway? What are the specific challenges with regard to (affordable) land and housing?

How would you characterize the peri-urban areas along Thika highway with regard to spatial segregation?

In how far can problems or challenges that result from the project be tackled in the future? Why might some of them be irreversible (in the short/medium term)?

What other factors or events (post-election violence, housing speculation) might have an impact on the current changes and development in the peri-urban areas along the highway?

Global vs. Local

What role does the THIP play for the Nairobi Metropolitan Region and for the local communities along the highway (housing, infrastructure, businesses, job market, social services)?

Would you see the THIP as a global project or a project that is directed to external aspects such as economic competiveness in East Africa?

In as far is the THIP a project of the metropolitan region? What stake does Nairobi, Ruiru, Thika, or other towns have in the project? What does Nairobi offer Ruiru und Thika in exchange for the benefits from the THIP?

What overlapping or conflicting roles of the project can you identify?

How could the project cater to local, national, and global objectives?

What institutions would play an important role in this regard? Do these institutions have sufficient resources, capacities, and capabilities to fulfill their role? Would you describe their role and activities as rather proactive or reactive?

Comparability

How would you see the THIP in comparison to other transportation infrastructure developments in Kenya, East Africa, or other developing countries?

What similarities or differences can you identify?

Do you know countries or cities that have done better in similar transportation infrastructure projects?

What were the reference projects the THIP designers/ decision makers based their ideas, objectives, and expectations on?

Do you think that decision makers (and private contractors, politicians, and the research community) sufficiently understand and take into account the need of concerned citizens in such projects?

Conclusion

What is your outlook on the future of the THIP and the socio-economic development of the area along the highway in particular?

Why might the THIP not result in positive growth with regard to the economy? (trickle-down…)

What are major lessons you would draw from the THIP (from your perspective as a XXX)? (Should the project not have been implemented at all?)

If it is only feasible to upgrade one type of infrastructure after another, should transportation infrastructure follow other basic infrastructures or the other way around? Why?

What can we expect from similar transportation infrastructure projects currently under way or planned for Nairobi or other Kenyan cities?

What recommendations would you give for such transportation infrastructure project or the THIP in particular (from your perspective as a XXX)?

What could you do to improve that situation?

9.10 Annex J: Qualitative Structured Content Analysis Applied to Interview Material

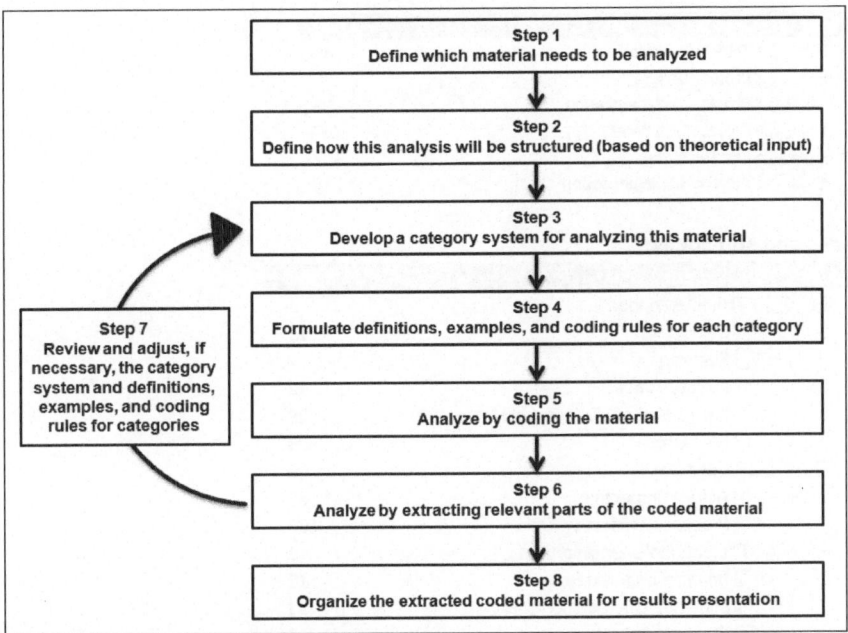

Qualitative Structured Content Analysis Applied to Interview Material (Source: Author translated/adapted from Mayring 2010: 93).

9.11 Annex K: MAXQDA – Code System

Summary:

Code System	1809
1 THIP Description	81
2 Project Design	176
3 Project Implementation	98
4 Project Outcomes	707
5 Perspectives	437
6 Beyond Implementation	310

Codes and sub-codes:

Code System	1809
1 THIP Description	0
Description	28
Flagship	7
History/roots	12
Kenya Vision 2030/Nairobi Metro 2030	29
NMT/uses	5
2 Project Design	0
Actor Perspective	67
Process Perspective	109
3 Project Implementation	0
Impacts on businesses	12
Land use changes/conflicts	13
Own role in implementation	12
Process + problems/reasons/solutions	33
Road reserve	28
4 Project Outcomes	0
Economic development	227
Political development	14
Provision development	103
Social development	135
Spatial development	116
Time Dimension	26
Winners/Losers	86
5 Perspectives	0
Historical development perspective	96
Institutional perspective	149
Political Arena perspective	48
Role perspective	76
Spatial perspective	68
6 Beyond Implementation	0
Comparability	58
Direction of project	44
Take home aspects	208

Complete code system:

Code System [1809]
- 1 THIP Description [0]
  - Description [28]
  - Flagship [7]
  - History/roots [12]
  - Kenya Vision 2030/Nairobi Metro 2030 [29]
  - NMT/uses [5]
- 2 Project Design [0]
  - Actor Perspective [0]
    - Actors + objectives/needs [39]
    - Conflict of interests/needs [10]
    - Opposing/missing actors + objectives/needs [18]
  - Process Perspective [1]
    - Information access, transparency, consultation [24]
    - Integrated project approach [47]
    - Process + reasons [22]
    - Top-down/bottom-up [15]
- 3 Project Implementation [0]
  - Impacts on businesses [12]
  - Land use changes/conflicts [13]
  - Own role in implementation [12]
  - Process + problems/reasons/solutions [33]
  - Road reserve [28]
- 4 Project Outcomes [0]
  - Economic development [0]
    - Agriculture [27]
    - Business and job opportunities [57]
    - Informal businesses [18]
    - Speculation, land/housing rates, financing [71]
    - Trade and regional connectivity [14]
    - Transport modes/flows [40]
  - Political development [2]
    - Public debate on THIP [6]
    - Relation b/w counties/regions [6]
  - Provision development [0]
    - Infrastructures/basic services [48]
    - Road design/safety [34]
    - Social services [21]
  - Social development [0]
    - Access [28]
    - Affordability/expenditures [30]
    - Community/liveability/environment/resources [30]
    - Displacement [32]
    - Segregation [15]
  - Spatial development [0]
    - Peri-urban areas [35]
    - Estate/slum phenomenon [16]
    - New housing developments [24]
    - Type of land uses/changes [41]
  - Time Dimension [0]
    - Immediate Outcomes [7]
    - Longer-term outcomes [9]
    - Problem tackling/irreversibility aspect [10]
  - Winners/Losers [9]
    - Losers + reasons [17]
    - Sharing benefits options [16]
    - Winners + reasons [44]
- 5 Perspectives [0]
  - Historical development perspective [0]
    - Economic-infrastructure development argument [30]

> Kenya's past and character [55]
> Post-election violence/elections [11]
> Institutional perspective [0]
>> Interrelation of planning [40]
>> Government Structure/New Constitution [38]
>> Quality/attitude of institutions/actors [38]
>> Implementation tools [33]
> Political Arena perspective [0]
>> Citizens' representation [15]
>> Decision makers' understanding [9]
>> Politics in Kenya [24]
> Role perspective [0]
>> Role of civil society [10]
>> Role of international actors [23]
>> Role of the private sector [18]
>> Role of the state [25]
> Spatial perspective [0]
>> City of Nairobi [12]
>> Githurai [6]
>> Juja Town [3]
>> Nairobi Metropolitan Region [27]
>> Other towns/metro regions [4]
>> Ruiru Town [2]
>> Satellite towns [5]
>> Thika Town [9]
6 Beyond Implementation [0]
> Comparability [0]
>> Idea origins/benchmarking [15]
>> Other countries [17]
>> Other road projects in Kenya [26]
> Direction of project [0]
>> Conflicts of scales [2]
>> Targeted scales [42]
> Take home aspects [0]
>> Conclusions + recommendations [70]
>> Infrastructure priorities [12]
>> Learning effects [28]
>> Outlook + threats [64]
>> Own role for improvement [10]
>> Symbolism/metaphor [24]

9.12   Annex L: The Quality of Qualitative Research

Part of a proper research process is the reflection on the quality of the conducted research. While quantitative research has long established codes of controlling for the quality of applied methods, the discussion on quality criteria in qualitative research has reached no clear consensus (Baxter & Eyles 1997: 505). While extensive textbooks exist on general quality criteria (Mayring 2010: 116-118; Helfferich 2011: 154-157; Lamnek 2010: 127-167), the analysis by very few authors, such as Baxter and Eyles (1997), shows how qualitative research has difficulties to meet quality standards and that these standards are difficult to define precisely. In this thesis, three strategies were developed to ensure the quality of this qualitative research work.

The first strategy was to follow well-established (German) textbook guidelines and check the material and the research process against the quality criteria of reliability, validity, and objectivity (for instance Mayring 2010: 116-118). Reliability describes how research findings and conclusions go beyond a single specific situation and could also be found in similar instances. Validity refers to the applied methodological approaches that are effective to produce research findings and conclusions for the formulated research questions and hypotheses. Objectivity describes how research findings and conclusions are independent of the researcher and the context in which they were produced. It becomes clear that qualitative research can never fulfill these three criteria to full extent. Nevertheless, a reflection on and an application of these quality criteria against the conducted research helped to understand context dependencies or comparability of empirical findings and conclusions in this research project.

This aspect is directly related to the second strategy: The research process was made transparent to the reader – which the methodological chapter of this thesis particularly aims at. This openness also includes a reflection on the complexity of the research object that denies definite, complete conclusions and calls for further research.

The third strategy to safeguard against one-sided, imbalanced research findings was the triangulation. Of the various elements in the triangulation strategy (Flick 2012: 310-315; Lamnek 2010: 245-265), the use of different sources and types of data, the reference to different theoretical research fields, and the use of several methodological approaches was applied. The idea behind triangulation is the development of a broader, more comprehensive understanding of the research object, without aiming for absolute completeness or prioritizing one method over the other (Mayring 2002: 147-148; Lamnek 2010: 142-143). As the remarks on additional methodological approaches in chapter 4.2 exemplify, it is the reflection on strengths and weaknesses of the applied tools that enriches the research – not the drawing of an over-loaded picture of the research object.

9.13 Annex M: Photos of the THIP Implementation Phase

Demolition of the Thika Road Nakumatt store (Source: Joseph Mathenge 2008, http://www.nation.co.ke/image/view/-/486200/highRes/48914/-/maxw/600/-/ebmrj8/-/Nakumatt.jpg).
Demolition of the Thika Road Nakumatt store (Kainvestor 2008, http://4.bp.blogspot.com/hZhftjtqaL0/ST5Ivq71A5I/AAAAAAAAAPY/J5BmjP8kJn4/s320/Nakumatt%2Bdemolition.jpg).

Demolitions of buildings along Thika Road (Source: Stephen Mudiari 2008, http://thma02.yimg.com/nimage/09cdd6df0f3715e6).
Road diversion during construction (Source: Liz Muthoni 2011, http://www.businessdailyafrica.com/image/view/-/1142722/medRes/252536/-/maxw/600/-/ohcejnz/-/thika%2Broad.jpg).

Vandalized power meter box (Charles Kimani 2012, http://www.the-star.co.ke/sites/default/files/styles/node_article/public/images/articles/2012/11/06/94304/picstory-thikaroad_0.png).
Road accident during construction (CSUD 2012, http://csud.ei.columbia.edu/files/2012/04/Thika-Road-Study-300x233.jpg).

Expansion works during rush hour traffic jam (Bert Yates 2010, http://2.bp.blogspot.com/_jJnAVqDkQKo/S4fbhCuw-YI/AAAAAAAAH38/d0_exoTOKig/s1600-h/...and+continues.....jpg).
Excavation works for an underpass of Thika Highway (Thika Road Blog 2011, https://lh3.googleusercontent.com/-Vq65to2u96E/TXYHp3axlhI/AAAAAAAAA00/w2sNwyyH3R8/s1600/Construction.jpg)

9.14 Annex N: Overview of THIP's Outcomes

The following two tables summarize outcomes of the THIP along the discussed issues in the empirical analysis. These outcomes are categorized as positive, ambivalent, or negative; although such categorization shall be treated with caution and read in context of the argumentation in chapter 5.5.

Changes Related to Transport and Economy

Traffic, Transport, Trade		
Positive Outcomes	*Ambivalent Outcomes*	*Negative Outcomes*
- Savings in travel time - Increased predictability of travel times - Decreased transport costs - Improved mobility for commuting and doing business	- Bottleneck problem at entrance into Nairobi - Increased pedestrian safety through footbridges in some parts, while large distances of footbridges in many parts leads pedestrians jumping onto the highway - Functioning road network still necessary (feeder and access roads) - Need for maintenance regime, particularly due to vandalism - Inter-regional trade improvements dependent on other road projects	- Insufficient improvements for public transit system and non-motorized transport - Accidents due to high-speeds and insufficient road safety design

Business and Job Opportunities		
Positive Outcomes	*Ambivalent Outcomes*	*Negative Outcomes*
- Increased economic vibrancy - New investments by companies - New job opportunities - More middle-income customers - Improved supply of goods and services	- Formal and informal businesses benefiting from growth nodes, while competition might increase - No formalization or integration of informal businesses	- Some businesses closed off from highway due to its limited access design - No demarcated market spaces for informal businesses - Displacement of informal businesses into areas further away from highway

Agriculture		
Positive Outcomes	*Ambivalent Outcomes*	*Negative Outcomes*
- Savings in travel time - Better accessibility to agricultural land and markets	- Conversion of agricultural land	- Balancing off of cost savings for farm produce due to increased distances of remaining agricultural land - Further aggravation of food security in the region - Loss of valuable arable land

Changes Related to Land and Housing

Land-Use Changes		
Positive Outcomes	*Ambivalent Outcomes*	*Negative Outcomes*
- Opening up of land	- Rapid land-use changes from agricultural to commercial and residential	- Scattered land development - Challenges for the ecological and political-organizational sustainability of the region

Infrastructure and Basic Services		
Positive Outcomes	*Ambivalent Outcomes*	*Negative Outcomes*
	- Small-scale, developer-driven solutions - Dependence on trunk infrastructure systems to be upgraded or provided - Lack of integration of the road infrastructure with other infrastructure - Lack of public sector guidance in opening up land through infrastructure provision	- Shortages in basic services provision due to increased demands - Illegal uses of infrastructure and basic services - Stress on ecosystem services - Supply of infrastructure and basic services based on socio-economic grounds - Increased land and housing prices due to developer-provided infrastructure and basic services

Land and Housing Markets		
Positive Outcomes	*Ambivalent Outcomes*	*Negative Outcomes*
- Increased market vibrancy - Non-profit area becoming investment zone	- Skyrocketing land and housing prices - Very strong free capitalist market economy effects - Prioritization of mobility over accessibility	- Speculative land holding in an imperfect market - Spatial segregation

Affordable Housing		
Positive Outcomes	*Ambivalent Outcomes*	*Negative Outcomes*
	- Balancing off of transport cost savings due to increased distances after displacement - Different opinions on affordability of land and housing in that area	- Steep price increases in land and housing - Displacement on socio-economic grounds

Upper-Market Housing		
Positive Outcomes	*Ambivalent Outcomes*	*Negative Outcomes*
- Availability of land and housing for middle-income people with decent commuting conditions	- Many projects in the exclusive upper-market range - Furthering of a particular lifestyle	- Real-estate/slum phenomenon

Social Services and Community Life		
Positive Outcomes	*Ambivalent Outcomes*	*Negative Outcomes*
	- Improvements in social services limited to private sector - Changes to community life due to new residents	- Disturbance of community interaction due to lack of context-sensitive highway design

9.15 Annex O: Spatial Dynamics of Agriculture in the Peri-Urban Area

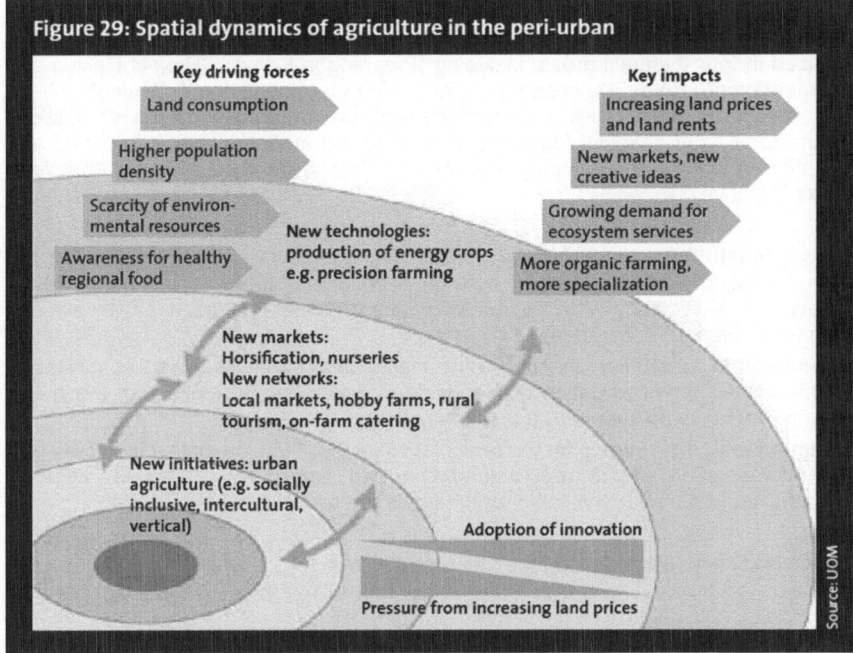

Figure 29: Spatial dynamics of agriculture in the peri-urban

Key driving forces

Land consumption

Higher population density

Scarcity of environmental resources

Awareness for healthy regional food

New technologies: production of energy crops e.g. precision farming

Key impacts

Increasing land prices and land rents

New markets, new creative ideas

Growing demand for ecosystem services

More organic farming, more specialization

New markets: Horsification, nurseries
New networks: Local markets, hobby farms, rural tourism, on-farm catering

New initiatives: urban agriculture (e.g. socially inclusive, intercultural, vertical)

Adoption of innovation

Pressure from increasing land prices

Source: UOM

Spatial dynamics of agriculture in the peri-urban area (Source: Piorr et al. 2011: 65).

9.16 Annex P: Overview of Policy Recommendations

The following list of policy recommendations was developed from both a literature research and the conducted qualitative interviews with experts in Kenya. They are organized in four thematic areas, including "Planning in Kenya", "Project Design and Implementation", "Land Development", and "Broader Institutional Setting". These policy recommendations are assessed against the three dimensions of political, organizational, and financial feasibility, based on the author's work experience and perception of politics and planning in Kenya as well critical discussions with several experts in Kenya.

Political feasibility: stands for the political will that is required to implement the recommendation and considers in how far a recommendation could be politically controversial, particularly with respect to electoral politics or relations with external actors (such as donor countries).

Organizational feasibility: stands for the institutional capacities and capabilities to implement the recommendation and considers in how far the culture of the public sector in Kenya would support the implementation.

Financial feasibility: stands for the financial costs and benefits that would come with the implementation of the recommendation and considers in how far financial resources could be made available for the implementation.

The feasibility is assessed along the following scale:
++ = very easy / + = rather easy / o = possible / − = rather difficult / − − = very difficult

Thematic Area	Recommendation	Political Feasibility	Organizational Feasibility	Financial Feasibility
Planning in Kenya	Government shall develop a culture of horizontal and vertical information sharing, coordination, and cooperation.	o	−	++
	Government shall further improve the capacity development of public officials.	++	o	−
	Government shall move from structural plans to strategic plans, from punitive measures to incentives, from stark regulation to guidance.	o	−	+
Project Design and Implementation	Government shall make infrastructure projects context-sensitive and grounded in specific micro-scale conditions with regard to accessibility aspects under the consideration of different design options and project alternatives.	o	o	o
	Government shall enable meaningful stakeholder engagement from start to finish based on transparent communication and accessible, easily understandable information.	−	o	+
	Government shall specifically include in project plans evaluations on a project's regional impacts, life-cycle costs, and maintenance necessities, with the formulation and application of clear mitigation measures towards identified repercussions.	o	o	+